NATO ASI Series

Advanced Science Institutes Series

A series presenting the results of activities sponsored by the NATO Science Committee, which aims at the dissemination of advanced scientific and technological knowledge, with a view to strengthening links between scientific communities.

The Series is published by an international board of publishers in conjunction with the NATO Scientific Affairs Division

A **Life Sciences** B **Physics**	Plenum Publishing Corporation London and New York
C **Mathematical and Physical Sciences** D **Behavioural and Social Sciences** E **Applied Sciences**	Kluwer Academic Publishers Dordrecht, Boston and London
F **Computer and Systems Sciences** G **Ecological Sciences** H **Cell Biology** I **Global Environmental Change**	Springer-Verlag Berlin Heidelberg New York London Paris Tokyo Hong Kong Barcelona Budapest

PARTNERSHIP SUB-SERIES

1. **Disarmament Technologies**	Kluwer Academic Publishers
2. **Environment**	Springer-Verlag
3. **High Technology**	Kluwer Academic Publishers
4. **Science and Technology Policy**	Kluwer Academic Publishers
5. **Computer Networking**	Kluwer Academic Publishers

The Partnership Sub-Series incorporates activities undertaken in collaboration with NATO's Cooperation Partners, the countries of the CIS and Central and Eastern Europe, in Priority Areas of concern to those countries.

NATO-PCO DATABASE

The electronic index to the NATO ASI Series provides full bibliographical references (with keywords and/or abstracts) to about 50000 contributions from international scientists published in all sections of the NATO ASI Series. Access to the NATO-PCO DATABASE compiled by the NATO Publication Coordination Office is possible in two ways:

- via online FILE 128 (NATO-PCO DATABASE) hosted by ESRIN, Via Galileo Galilei, I-00044 Frascati, Italy.
- via CD-ROM "NATO Science & Technology Disk" with user-friendly retrieval software in English, French and German (© WTV GmbH and DATAWARE Technologies Inc. 1992).

The CD-ROM can be ordered through any member of the Board of Publishers or through NATO PCO, Overijse, Belgium.

Series I: Global Environmental Change, Vol. 35

Springer
Berlin
Heidelberg
New York
Barcelona
Budapest
Hong Kong
London
Milan
Paris
Santa Clara
Singapore
Tokyo

Clouds, Chemistry and Climate

Edited by

Paul J. Crutzen
Max-Planck-Institute for Chemistry
Postfach 3060
D-55020 Mainz, Germany

Veerabhadran Ramanathan
Center for Clouds, Chemistry, and Climate
Scripps Institution of Oceanography
La Jolla, CA 92093-0239, USA

Springer
Published in cooperation with NATO Scientific Affairs Division

Proceedings of the NATO Advanced Research Workshop "Clouds, Chemistry and Climate", held at Ringberg, Germany, March 21–25, 1994

Cataloging-in-Publication Data applied for

Die Deutsche Bibliothek - CIP-Einheitsaufnahme

Clouds, chemistry and climate : [proceedings of the NATO Advanced Research Workshop "Clouds, Chemistry and Climate", held at Ringberg, Germany, March 21 - 25, 1994] / ed. by Paul J. Crutzen ; Veerabhadran Ramanathan. - Berlin ; Heidelberg ; New York ; Barcelona ; Budapest ; Hong Kong ; London ; Milan ; Paris ; Santa Clara ; Singapore ; Tokyo : Springer, 1995
(NATO ASI series : Ser. I, Global environmental change ; Vol. 35)
Additional material to this book can be downloaded from http://extra.springer.c

ISBN-13:978-3-642-64672-0 **e-ISBN-13:978-3-642-61051-6**
DOI:10.1007/978-3-642-61051-6

NE: Crutzen, Paul J. [Hrsg.]; Advanced Research Workshop Clouds, Chemistry and Climate <1994, Ringberg, Kreuth>; NATO: NATO ASI series / I

ISBN-13:978-3-642-64672-0

Softcover reprint of the hardcover 1st edition 1996

Typesetting: Camera ready by authors/editors
Printed on acid-free paper
SPIN: 10466185 31/3137 - 5 4 3 2 1 0

FOREWORD

The objective of this NATO Advanced Research Workshop was to discuss our current understanding of the role of clouds in climate and chemistry. The range of topics discussed during the workshop included: modeling of clouds in GCMs; observations of the cloud micro physical properties; the water vapor cycle; troposphere-stratosphere exchange; role of in-cloud transport in tropospheric ozone; regulation of current and paleo climate by clouds; and anthropogenic sulfate aerosols and modification of cloud properties. The essence of the discussions is captured in the accompanying summary by the rapporteurs and the chapters by some of the speakers.

The underlying message is that significant progress has been made, resulting in exciting new developments in our perception of the role of clouds in the global system.

* The tropical convective-cirrus cloud systems emerge as a major influence on the climate system. Micro physical processes, such as the rate of precipitation and re evaporation of ice particles, seem to regulate the large scale vertical distribution of water vapor which is the dominant greenhouse gàs. Water vapor data collected during the Central Equatorial Pacific Experiment (CEPEX), document the large moistening effect of deep convection on scales of thousands of kilometers.

A major chemical finding in the same region was the observation of extremely low ozone volume mixing ratios of less than 10^{-8} in the entire troposphere of the central equatorial Pacific extending over a distance of about 2000 km. This finding establishes the Pacific as a major chemical sink region for tropospheric ozone.

It is well known that ozone can be destroyed efficiently in the marine boundary layer by the reactions

$$O_3 + h\upsilon \rightarrow O(^1D) + O_2 \quad \lambda \leq 325\ \text{nm}$$
$$O(^1D) + H_2O \rightarrow 2OH$$

However, is is not yet clear whether these reactions alone are sufficient to explain the large ozone depletion (Kley et al., 1995) or if additional chemical ozone loss may take place through aqueous -or heterogeneous- phase processes in water -or ice- clouds, respectively (Crutzen, 1995). The very low ozone mixing ratios, in conjunction with the high albedo of deep convective clouds, may cause suppressed hydroxyl concentrations. If so, the transport of sulfur- and halogen- containing organics to the upper troposphere and, perhaps, even to the lower stratosphere could be more efficient than hitherto thought which might influence sulfate particulate (CCN) and ozone depletion reactions, e.g. via iodin catalysis (Solomon et al., 1995).

Data collected during CEPEX demonstrate that the radiative energy converging into the convective-cirrus systems is a significant source for the diabatic heating of the upper troposphere. GCM studies reveal that this cirrus heating contributes significantly to the tropical Walker and Hadley circulation.

Until recently, we were unable to measure the micro physical characteristics of these cirrus clouds. Breakthroughs in instruments (CVI instrument) and availability of high altitude aircraft (ER-2) are enabling us to have a glimpse of this important issue. Preliminary data seem to show the existence of numerous small crystals (size $\leq 20\ \mu m$), which have to be included in models. But the CEPEX data also show the preponderance of large crystals ($\geq 250\ \mu m$), which have also been ignored in models.

The CEPEX data demonstrated that the reduction of solar radiation at the sea surface by the convective-cirrus clouds far exceeded their greenhouse effect at the surface, thus exerting a large net radiative cooling at the sea surface over the warm tropical oceans. It is still a matter of considerable debate whether this convective-cirrus cooling effect is the dominant regulator of the surface temperatures in the western Pacific warm pool or competing processes are in effect. The discussion in this volume, not only captures some of the essentials of this debate, but also reveals the primitive state of our understanding of this fundamental issue in climate and climate change. Tropical temperature changes during the last glacial period and warm epochs should offer clues regarding the stability of tropical climate. Studies of the GRIP and GISP2 ice cores demonstrate that large and rapid climate changes occurred during the last ice age and during the transition towards the present Holocene. Such changes are best documented in the North Atlantic and we need to document the past changes in the tropical oceans.

* The concept of stratospheric-tropospheric exchange has evolved over the last few years. This pertains particularly to extratropical wave disturbancies as drivers of the "overworld" meridional circulation with upward transport in the tropics and compensating downward transport at extratropical latitudes (see review by Holton and Lelieveld at this workshop). However, much more needs to be learned about the transport in and to the lowest parts of the stratosphere in the tropopause region. Observations during CEPEX in the tropics have shown record low H_2O volume mixing ratios of about 1 ppm which clearly indicates the importance of local, small scale, exchange processes for the trace gas composition of the lower stratosphere.

Recent observations by the Upper Atmospheric Research Satellite have been of particular significance for an understanding of stratospheric dynamics. With regard to the transfer of ozone from the stratosphere into the troposphere at extratropical latitudes, the correlations between NO_y and O_3, characteristic of air masses with origin in the lower "overworld" (Murphy et al.,1993), combined with the NO_y production rate by the reaction $N_2O + O(^1D) \rightarrow 2\ NO$ establish the role of O_3 and NO_y as conservative tracers in the lowest few kilometers of the stratosphere. This does not, however, applies to the late winter and early spring periods at high latitudes (especially over the Antarctic continent), probably implying a gradual decrease in the downward flow of ozone to the troposphere during the past two decades with some potential repercussions for tropospheric chemistry.

* The role of anthropogenic sulfates in climate is clearly an important issue for future climate changes. Scattering of solar radiation by sulfates enhance the clear sky albedo (the direct effect) and can also enhance the number of cloud drops, which in turn can enhance the cloudy sky albedo (the indirect effect). Together, the direct and indirect effect of sulfates during the last century may be as large as the combined greenhouse effect. But, unlike the sound observational basis for the increases in concentration of greenhouse gases, the sulfate effect awaits observational confirmation. The direct effect depends critically on the rate of formation of new particles through gas to particle conversion of the emitted SO_2. The indirect effect depends on the increase in cloud droplet number concentration in marine cloud systems. Recognizing the need for observational verification, inter national field campaigns have been initiated in the north Atlantic. But, the Far East and Asian countries are projected to be the major emitters of SO_2 in the next century, and the role of sulfates in the tropical setting should be investigated. In our discussions, the equatorial Indian ocean emerged as a potential site for a field campaign to test the sulfate hypothesis. The equatorial Indian ocean, during the spring season, is down wind of continental air from the Indian sub continent with very little precipitation along the way. Furthermore, the continental air ultimately converges into the inter tropical convergence zone, which provides a contrast between the polluted air from the north to the pristine air from the southern Indian ocean.

References

D. Kley, H.G.J. Smit, H. Vömel, S. Oltmans, H. Grassl, V. Ramanathan and P.J. Crutzen: Record Low Ozone Observatons in the Convective Regions of the Pacific, Science (submitted)

P.J. Crutzen: Ozone in the Troposphere, in "Composition, Chemistry and Climate of the Atmosphere" (H. Singh, Editor), pp. 349-393, Van Nostrand, 1995.

S. Solomon, R.R. Garcia and A.R. Ravishankara: On the Role of Iodine in Ozone Depletion, J. Geophys. Res. **99** (D10), 20491, 1994.

D.M. Murphy et al.: Reactive nitrogen and its correlation with ozone in the lower stratosphere and upper troposphere, J. Geophys. Res., **98**, 8751, 1993.

The Editors

ACKNOWLEDGMENT

The NATO Advanced Research Workshop on Clouds, Chemistry and Climate was supported by the NATO-ARW930752 grant and the NSF-ATM8920119 grant to the Center for Clouds, Chemistry and Climate.

Pre-press production of this manuscript was supported by the Vetlesen Foundation Grant to the Scripps Institution of Oceanography.

This is publication #144 of the Center for Clouds, Chemistry and Climate, a National Science Foundation Science and Technology Center.

TABLE OF CONTENTS

NATO ARW ON CLOUDS — SESSION SUMMARIES

Session 1: Radiative Convective Interactions in the Tropics
Rapporteur: T.E. Nordeng, Norwegian Meteorological Institute, Oslo, Norway

The two papers presented in this session focused on the role of dynamics in the water vapor distribution of the tropics and its implications to climate.

Emanuel addressed the problem of describing microphysics in parameterization of cumulus convection and its impact on climate simulations. He demonstrated with his scheme for cumulus convection that very different water vapour profiles are obtained in radiative convective equilibrium runs by varying tunable constants in the microphysical part of the scheme. He stressed the importance of water vapour as the by far mostimportant greenhouse gas on its own, but also indirectly by its role in generating clouds.

Emanuel also asserted that none of the existing schemes handle microphysics correctly. Instead of verifying schemes against observed mass flux and apparent heating, which according to Emanuel is an easy task because they are constrained by radiative cooling, one should assess the schemes against observed profiles of relative humidity.

Quite a lot of the general discussion during the whole workshop was related to the theory which Ramanathan and Collins presented in an article in NATURE (vol 351, p27, 1991) in which they proposed, based on observations in the Western Pacific region, that there is an upper limit to the sea surface temperatures because clouds act like a thermostat. Ramanathan and Collins found that the earth's greenhouse effect increases sharply as the temperatures increases beyond 300K and that it is the longwave absorption in clouds that are responsible for this. At the same time, however, cloud albedo increases as well, but at a faster rate than the greenhouse effect implying that there eventually will be an equilibrium which limits further increase of SST. This was challenged by R. Pierrehumbert, who stressed that it is not correct to look at the warm pool of the earth in isolation, but that the corresponding dry "cold" areas should be of equal importance. The dry air has a controlling effect. In his talk he presented various ways from which the dry air could be created, and observations were used to identify "tropospheric rivers" which ventilates the tropics with middle latitudes. He also presented observations from Eastern Pacific showing some extremely dry air. It turned out that the "dry pool" seemed to be generated locally, i.e. from large scale stationary subsidence.

The general discussion which followed these two papers were mostly related to the role of microphysics and large scale dynamics versus local equilibrium. It was mentioned that the terminology "microphysics" is misleading and have different meanings for cloud physicists and general circulation modellers. The possibility of using high resolution cloud models to study microphysical processes (and hence derive parameterization methods) was forwarded. It was also mentioned that our lack of knowledge about microphysical processes and particularly how they should be represented in GCM's are very discomforting.

The role of large scale dynamics versus local equilibrium was discussed in relation to Pierrehumbert's paper. It was mentioned that the convective activity in the Western Pacific is a composite of several convective episodes and that it is important for climate modelling to get the transient activity right.

Session 2: Role of Cirrus Clouds

Rapporteurs: S. Borrmann[1] and A. Heymsfield[2]
[1]Institut für Physik der Atmosphäre, Universität Mainz, D-55099 Mainz, Germany
[2]National Center for Atmospheric Reasearch, Boulder, CO 80307, USA

The three presentations in this session were concerned with the microphysical and optical properties of cirrus cloud particles and their influence on radiative transfer calculations.

The paper by Y. Fouquart addressed three main questions: (1) What are the differences in the optical properties between cirrus clouds and liquid water clouds? (2) How are the radiative transfer calculations influenced by the inclusion of non-spherical particles?, and (3) What are the possibilities and limitations of simple parameterizations of radiative transfer treatment in cirrus clouds. The sensitivity of climate models depends on crystal size and shape to the degree that even the sign of the cloud water temperature feedback is uncertain. Another major problem is the spatial and the temporal variability of cirrus cloud particle sizes and shapes. These variations may become so large that it is difficult to generalize results obtained from a few measurements because cirrus are horizontally inhomogeneous on scales of one kilometer. Addressing the first question, Fouquart presented evidence to show that ice clouds cause much more side scattering than water droplets. As a consequence of these differences the bidirectional reflectance functions and angular variation of polarization rate differ greatly between water droplets and cirrus crystals. Also it was mentioned that there is evidence for preferential orientation of the cirrus cloud crystals. If this is the case then the scattering functions and extinction depend on the direction of the incident light. With respect to the second question, the differences in the asymmetry factors for the same equivalent mass were shown for plates, columns and spheres. Based on theoretical calculations the reflectance of cirrus clouds may be much larger than that for the water droplet clouds. Simple radiative equilibrium models show that cirrus consisting of crystals with asymmetry factors of $g=0.87$ may increase the surface temperature twice as much as cirrus with g=0.7 crystals. Regarding the third question, the following methods for calculation of single scattering parameters were discussed: Mie (1908) electromagnetic theory; Liou (1972) and Stevens (1979) for infinite cylinders; Asano and Sato (1983) for spheroidal particles; discrete dipole approximation theory (e.g. Flatau, 1988); anomalous diffraction theory for large size parameters (Van de Hulst, 1957) and various geometric ray tracing methods for hexagonal ice crystals (e.g. Jacobowitz, 1971).

The paper of H. Pruppacher elaborated on the severe discrepancies for observed ice nucleation rates and those predicted by classical nucleation theory at the cirrus cloud level. The most important parameters for water appearing in the nucleation rate equation are: specific and latent heat, density, viscosity, surface tension, self diffusion of the water molecules, activation energy for self diffusion and viscous flow and the interface energy between water and ice. He showed that the discrepancies can be resolved if the previously adopted values for these parameters are replaced with data published in the physico-chemical literature during the 1980's. The previously used data were determined by extrapolation from measurements at 0° C while the more recent values were measured down to -33° C by means of new techniques and further extrapolated beyond -40° C. These data show that supercooled water exhibits a singularity at -45° C. The repeatedly occurring limit of -45° C in the properties of water can be explained in terms of a cooperative process among the hydrogen bonds of the water molecules. This leads to the formation of water clusters which become increasingly larger as the temperature is decreased. As a result, the larger cluster instead of single water molecules will transfer across a growing water/ ice interface. These clusters will have breaking hydrogen bonds at their periphery and a continuous decrease of the molar activation energy for transfer across the ice/water interface can be expected. This activation energy cannot be measured. For indirect determination the values for the nucleation rate observed at the cirrus cloud level and in cloud chambers were verified by comparing them with the lowest (laboratory) temperatures to which single droplets could be supercooled. Subsequently these values were introduced into the nucleation equation together with the newly available data for the parameters involved and the equation was solved for the energy of transfer which is the only remaining unknown. As expected from the cluster behavior of supercooled water, the activation energy for transfer was found to decrease with decreasing temperature. Thus a physically consistent model is now available for computing the temperature at which the water drops of any given size will freeze by homogeneous nucleation in any given environment.

J. Heintzenberg reported on his in-situ measurements of small (i.e. submicron size) cirrus crystals. As the cirrus cloud life cycle commences and ends with submicron particles, these have to be taken into account for radiative transfer calculations. Submicron H_2SO_4

droplets or particles coated with H_2SO_4 grow at cirrus levels as liquid phase by H_2O vapor uptake until they are sufficiently dilute to freeze. Once frozen they grow as crystals. In previous in-situ measurements particles smaller than 20mm were not measured despite increasing number size distributions at the low particle size end of the distribution. To detect submicron cirrus crystals in-situ, an airborne counterflow virtual impactor (CVI) was flown on a Falcon aircraft during ICE 89, Eucrex 92 and 94 over the North Sea, the Alps and Germany. The CVI detects particles from 5mm to 50mm Stokes diameter. Additionally condensation nuclei, H_2O vapour etc. were measured. The crystal size distributions were inferred from the measured number of crystal residue particles and the cloud water content. The CVI frequently detected particles when the PMS optical particle counters showed zero particle counts. The cloud water content in the CVI detection range was found to be of the same order as that for particles larger than 50mm. Up to 3000 particles per liter (i.e. a very high number) were measured by the CVI. Persistent cirrus were found 500-1000 km downwind from their likely place of generation. Heintzenberg pointed out three open questions/problems: (1) The size distribution of the crystal residues did not resemble the aerosol size distributions measured with the same sensor (2) In air sub-saturated with respect to ice, the logarithm of the crystal number concentration was well correlated with air temperature and (3) Well below 100% r.h. with respect to ice, there were geographically stationary cloud structures of one hour time scales which could be recognized in a concurrent IR satellite image. Although not detected at the beginning by the PMS probes, these structures were also seen at the end of the flight by PMS sensors.

The major hot topic related to cirrus crystals was the need for detailed measurements of the phase functions of different crystal types and sizes. Also the theoretical methods need to be refined for using the calculated phase functions in radiative transfer models. The necessity of techniques (preferably remote sensing) was emphasized in order to obtain a better data basis on the variations of temporal and spatial cirrus crystal distributions.

Session 3: Paleoclimate

Rapporteur: S.E. Schwartz, Brookhaven National Lab., Upton Long Island, NY 11973, USA

Hammer and Jouzel presented recent research on climate change in paleogeological times, focusing on results form ice cores in Greenland (Hammer) and the Vostok core in Antarctica (Jouzel), but also drawing on information from ocean sediment cores and other paleoclimate sources. The ice cores provide a richly detailed record of climate and climate change over a period extending back to 200,000 years before present (BP). Annual layer counting is used to precisely date the two Greenland cores down to ca 40 kyr BP. For earlier periods an age model is used. At Vostok the age model is used from the surface because the accumulation is too low for the annual layers to be counted. For Greenland cores, there has been discussion of possible errors due to folds or other disruptions associated with ice flow for the deepest layer especially over the previous interglacial (Gemian). The GRIP camp at the center of the Greenland dome is favorably located from this perspective (relating to present ice flow).

Principal features in ice cores that are examined are isotopic abundances of $^{18}0$ and D in water, which vary with temperature because of Rayleigh distillation, and concentrations of chemicals especially H^+ (acidity), $S0_4^{2-}$, NO_3^-, methanesulfonic acid (MSA) and dust. Large spikes in $S0_4^{2-}$ and H^+ can be related to known volcanoes in historic times, often exhibiting widespread geographical distribution. Spikes of volcanic ash are also observed but these are generally more local, e.g., due to an Icelandic volcano, observed only in Greenland, perhaps just in a small area of the ice sheet. Dust is eolian in origin, with principal mass in the size range 0.4-4μm radius.

A key feature noted in the Greenland ice record is the occurrence of numerous periods of less glaciation (interstadials) occurring during the last glacial period. These interstadials last roughly 500 to 2000 years, with a rapid warming towards temperature intermediate between interglacial and full glacial, followed by a slower cooling and a relatively rapid return to glacial conditions. The time record is characterized by its complexity. It is triggered in part by oscillating changes in insolation due to astronomical causes ("Milankovitch forcing"), but also presents more erratic features, which are attributed to the highly nonlinear nature of the climate system and may also correspond to purely internal instabilities of the system. There is a general asymmetry between cooling episodes (which often happen slower) and warming episodes (often more rapid). Some of the climate transitions may be rapid or very rapid. Here rapid means within the span of a human lifetime and very rapid means two years to perhaps half a year in some quantities such as the rate of dust accumulation. These changes are geographically widespread (north polar to south polar) although the rapidity of the change sometimes seen in the Greenland record (0.5-2 years) is sometimes not matched in Antarctic record where the changes are less important and much less abrupt. Recent research on the Greenland cores show remarkably rapid high magnitude fluctuations in the Eemian period (113-135 kyr BP) and instances of interglacial periods with temperatures somewhat warmer than at present. However questions have recently arisen regarding the validity of interpreting those isotopic changes in terms of climate changes because of differences between the two summit cores for this part of the record. Key observables exhibiting change between glacial and interstadal epochs are isotopic abundances of D and $^{18}0$ and concentrations of MSA, dust and trapped gases - CO_2 and CH_4. Dust exhibits much lower concentrations (order of magnitude) in interstadal periods than glacial. This can be due only in small part to the rate of water deposition (lower in glacial epochs by a factor of 2) and is attributed to greater source strength resulting from larger exposed area of soil surface (resulting from lower sea level), and possibly drier soils (resulting from decreased precipitation) and possibly greater wind speeds. CH_4 concentration is high in interstadal periods, possibly because of greater production rate in moist soils.

In the discussion it was noted that the climate change events noted in the paleo record present a challenge to the climate modeling community. In particular with reference to the present workshop on clouds, chemistry and climate there are issues where this research community may be able to provide insights or where present understanding may be challenged. It was observed that the same laws of microphysics must continue to apply, despite differences in the microphysical climate between glacial epochs and the present. Atmospheric chemists may be able to contribute insight to interpretation of relations between deposition rates of substances and their atmospheric concentrations. A key issue to the ice core community is reconstruction of the water deposition flux and more generally improved understanding of the hydrological cycle in glacial epochs. Did the decrease in precipitation extend to lower latitudes, as perhaps interpreted in lower CH_4 concentrations in glacial epochs?

An issue that arose in discussion was the relation between temperature and isotopic composition. In both polar regions an excellent correlation is observed for example between temperature and $\partial^{18}0$ that is used to infer temperature from the measured $\partial^{18}0$. However it was noted that the isotopic composition reflects the lagrangian history of the water deposited in precipitation, and thus the empirical calibration is perhaps subject to question if this lagrangian history were systematically different in climatic epochs different from the present. Attempts to model changes in $\partial^{18}0$ using GCM's have thus far not been entirely successful. Perhaps the key question facing the paleoclimate community is the reason for the complex and nonlinear nature of the climate record and, closely coupled, the mechanism whereby the transition can be as rapid as observed. It was noted that the transitions are not symmetric; that warming can take place much more rapidly than cooling. A suggested mechanism involves marginal ice-sheet instability with resultant massive amounts of fresh water ice entering the ocean and, upon melting, leaving low density pools of surface water that inhibit large scale ocean circulation, perhaps affecting the so-called "conveyer belt" circulation. There was considerable discussion of the possible role of dust as a climate forcing agent, thus leading to a possible feedback. Dust in clouds can enhance short-wave absorption in clouds thus decreasing solar heating at the surface. Changes in cloud microphysical properties due to dust may also contribute to changes in the hydrological cycle. The ice core measurements of dust (and perhaps also sulfate) can possibly be used in cloud models to address this issue.

Session 4: Modeling and Parameterizations

Rapporteurs: K. A. Emanuel[1] and J. Lelieveld[2]
[1]Massachusetts Institute of Technology, Cambridge, MA 02139
[2]Wageningen Agricultural University, Wageningen, The Netherlands

The papers presented in this session all illustrate the trend towards explicit prediction of cloud water content in global climate models in an effort to improve the representation of cloud-radiation interaction. Experiments with GCM's indicate a significant sensitivity of climate to the way in which this is done.

A major step toward making it possible to go about this in a reasonable way has been the advent of ERBE data, which allows a direct comparison of model-derived short- and long-wave fluxes at the top of the atmosphere with observations. The paper presented by E. Roeckner illustrates that even with very simple physics of cloud water and ice formation and destruction, one can obtain quite good agreement between model-derived and observed radiative fluxes at the top of the atmosphere, especially in the zonal averages of these quantities. That the sum of the net shortwave and longwave cloud effects should be well represented is not too surprising, since this is often a small net contribution, especially in the tropics. However, longwave cloud forcing in the storm track regions appears to be underestimated. Nevertheless, the improvement in the agreement between calculated and observed longwave and shortwave contributions in the latest generation of GCM's is indeed encouraging. We note, however, that far better observations of the terms in the surface energy budget will be necessary for complete validation of the model representations of cloud-radiation interaction.

Sensitivity to cloud-radiation interaction affects climate directly through the net heat budget of the planet and indirectly by changing circulation patterns owing to the inhomogeneity of cloud cover. P. Rasch demonstrated that merely accounting for the differing optical properties of continental and maritime clouds, owing to the greater number and smaller size of continental condensation nuclei, thus increasing the cloud reflectivity, can affect the extratropical 500-mb height pattern by as much as 10 dm. Moreover, the decreased solar flux to the surface over the continents reduces evaporation and consequently precipitation rates. In general, the distinction between continental and maritime cloud droplet numbers and sizes has improved the model performance. On the other hand, we are skeptical of some reports of sensitivity of tropical circulation to cloud optical properties based on simulations with fixed ocean temperature. Thermally direct circulations in the tropics maintain horizontal temperature gradients at values close to the critical gradient necessary for such circulations, and the adjustment of sea surface temperature is an important part of this process. Changing an internal atmospheric property, such as cloud emissivity, while forbidding changes in sea surface temperature may lead to very large changes in tropical circulation, as demonstrated by Roeckner; such large changes would probably not occur in models in which the sea surface temperature can respond to changes in the heat budget.

One issue that arises in this context is the relative importance of further refinements in accounting for cloud optical effects versus improving other aspects of climate models such as ocean circulation and water vapor control. It may be argued that the answer to this question will not be obtained until we have full confidence in our handling of cloud-radiation interaction. One process that continues to be of great concern in climate modeling is the representation of moist convection. T-E. Nordeng reminded us that the circulation in high resolution models (e.g., that of the ECMWF) is sensitive to the way in which convection is handled. He described an extension of the ECMWF cumulus convection scheme by parameterizations of organized entrainment and detrainment. The goal was to achieve closure between convective activity and buoyancy as an alternative to the moisture convergence closure used in the current operational model. In particular, the work of Thor-Erik Nordeng adds to the growing body of evidence that convection schemes controlled by water budgets, rather than buoyancy, are rather badly behaved.

In the discussion it was emphasized that cloud formation in GCM's is sensitive to both convective and microphysical parameterizations.

Session 5: Warm Pool SST Regulation: Primary Results from CEPEX
Rapporteur: W.D. Collins, Scripps Institution of Oceanography, La Jolla, CA 92093, USA

The range of sea surface temperatures in the tropical oceans has not changed significantly over the last 10000 years. Ramanathan and Collins (1991) (RC) introduced a "thermostat hypothesis" to explain two related issues for the present climate: first, why are maximum SSTs (Ts) so close to the 300K threshold for deep convection; and second, what are the restoring forces that limit Ts and deep convection? The thermostat feedbacks stabilize the climate in warm oceanic regions where the downwelling infra-red radiation increases faster than the upwelling thermal emission and where oceanic heat transports are small. The thermostat mechanism is based upon the shading of the surface by large cirrus anvils produced by tropical convective storms. Simply put, the thermostat hypothesis states that as Ts rises above a "convective threshold", nominally 300 K, convective storms and the associated extended cirrus clouds develop that substantially reduce the surface insolation. The reduction of solar radiation in conjunction with other processes cools the ocean and maintains the temperature close to the convective threshold.

1. CEPEX and related field observations
The Central Equatorial Pacific Experiment (CEPEX) was designed to provide observational evidence of this hypothesis. The CEPEX science team measured the gradients in

radiation and evaporation at the interface between the western Pacific warm pool and the colder eastern Pacific. The investigators also measured cloud microphysical parameters in order to understand the radiative effects of convective anvil clouds. These measurements, in combination with the TOGA Coupled Ocean-Atmosphere Response Experiment (TOGA/COARE), provide an extensive data set on the tropical climate system. Some of the preliminary results from CEPEX are discussed in this chapter. The basis of current CEPEX investigations is the heat budget of an ocean column and its change with Ts. However, a comprehensive investigation of thermostat mechanisms involves a combination of observations, simple empirical models, and coupled ocean-atmosphere GCMs. The GCMs are able to simulate the enormous range of time and length-scales and the nonlocal interactions beyond the reach of a field experiment. The observations from CEPEX and TOGA COARE will be useful for testing and validating the physics in these coupled models. In addition to satellite radiation and cloud imagery, the Tropical Atmosphere-Ocean Buoy system is an important on-going, long-term data base for CEPEX. Because of limited resources and the one-month time period, the CEPEX approach is to "ground truth" the satellite and buoy data with the aircraft and ship data where ever possible and then use those longer and more widely geographically distributed data for obtaining general conclusions.

2. Convection and large-scale dynamics

One of the most important open issues is the relation between Ts, the dynamics of convection, and the dynamics of large-scale circulation. The link between convection and large-scale circulation is a central component of the cirrus thermostat. Several modeling studies have suggested that shortwave and longwave flux convergence in tropical cirrus anvils may help accelerate the Walker circulation and efficiently redistribute diabatic heating over a large area. The basic dynamics of this redistribution are not well understood, and the effect has not been demonstrated from observations. It may be possible to study the response of the circulation to cirrus heating using simple two dimensional models. Increased convection over warmer oceans is associated with subsidence and suppressed convection elsewhere. One of the alternative thermostat mechanisms discussed at the workshop is based upon radiative cooling in the clear-sky subsidence regions rather than shortwave cloud forcing in the convective regions. The relative importance of convection and subsidence for the tropical energy budget is being investigated with CEPEX and TOGA COARE observations.

3. Ocean response

The preliminary CEPEX findings indicate that convection significantly reduces the net heat flux into the ocean surface. The effect on the sea-surface temperature depends on the interaction between atmospheric forcing and the response/feedback from the ocean mixed layer. In general the response of the mixed layer is slower than the forcing by atmospheric weather events, which last on the order of one to two days. However, there appears to be a range between several weeks to years where atmospheric and oceanic time scales are matched. The heat content of the mixed layer also depends on the removal of heat from the layer by oceanic advection and mixing. Previous observations suggest that the advective flux is less than 20 $W.m^{-2}$. This implies that the mixed layer heat budget is approximately equal to the net surface flux in the western Pacific warm pool. Results from TOGA COARE should help determine the magnitude of advection and the relation of mixing events to convection in much greater detail.

4. Greenhouse effect and cloud forcing

The CEPEX data shows that convection moistens the atmospheric column locally, and the increase in the surface greenhouse heating associated with convection has been measured for the first time. The convective clouds also heat the surface with longwave radiation, but this forcing is less than 20% of the longwave cloud forcing for the entire column. Studies employing ERBE, TOGA-COARE, and CEPEX data are suggesting that the reduction in insolation by cloud shading should be 50% larger than predicted by radiative models. The reduction is apparently caused by enhanced shortwave absorption in the cloudy tropical atmosphere. The net effect of convection on the local radiation balance is dominated by this cloud shading effect and results in a large reduction in the radiative heating of the sea surface. This cloud shading effect is much larger than the increase in surface greenhouse heating associated with deep convection. If this calculation of the absorption is verified with other observations, it will require a major revision of the treatment of cloud radiative transfer in GCMs.

5. The role of evaporative cooling

A strong, positive relation between evaporation and Ts has been proposed by many investigators as the principal mechanism limiting Ts. Preliminary CEPEX results indicate that evaporation east of the dateline increases with Ts, which seems to favor those arguments. However, simple mathematical models, more sophisticated numerical models, COARE evaporation measurements, and long-term estimates of evaporation from TAO buoys suggest that convection in the Pacific warm pool west of the dateline is associated with a decrease in evaporation as Ts continues to increase. Since CEPEX and COARE observations show the increased effect of cloud shading west of the dateline, the importance of the reduction in insolation relative to evaporation must be increasing rapidly. The connection between convection and evaporation will be studied further to understand the effect of convection on both the near-surface humidity and wind fields.

6. Significance for GCMs

The CEPEX observations suggest necessary conditions for coupled ocean-atmosphere models to guarantee physically accurate simulations of the tropical climate:

• The models must reproduce the growth in the frequency of convection with SST. The increase has been demonstrated using several independent long-term satellite records. The frequency begins to decrease with SST above 29.5C, but SSTs in this range are relatively rare.

• The long-term average wind speeds and evaporation for the warm pool from the simulations should be lower than the corresponding fields in the eastern Pacific. The probability distribution of wind speed and convection has been determined from moorings in the TOGA TAO array.

• The models must produce a super-greenhouse effect at the surface and at the tropopause. CEPEX observations confirm the enhanced greenhouse trapping detected by satellite, and have shown that the net radiative emission from the ocean decreases with increasing SST.

• The advection of heat out of the model mixed layer should be a small component of the layer heat budget if previous observations are confirmed by results from TOGA COARE.

• The treatment of cloud radiative transfer should reproduce the enhanced absorption of shortwave radiation in the cloudy atmosphere.

Improvement in the coupled GCMs will be critical for continued investigation of thermostat mechanisms for the tropical climate

Session 6: Stratosphere-Troposphere Exchange

Rapporteurs: P. Rasch[1] and Th. Peter[2]
[1]National Center for Atmospheric Research, Boulder, CO 80307, USA
[2]Max-Planck Institut für Chemie, D-6500 Mainz, Germany

J. Holton noted that our knowledge of stratosphere/troposphere exchange (STE) is in an impoverished state, and that STE is particularly important for understanding the distributions of chemically reactive species in the lower stratosphere and upper troposphere.

He noted that a particularly easy way of moving tracers across the tropopause was along isentropic surfaces. Some isentropic surfaces are contained only in the stratosphere. These define the "overworld", some are shared between stratosphere and troposphere, (the middleworld), and some reside only in the troposphere. The movement along isentropes in the middleworld provide an easy mechanism for STE.

The Brewer-Dobson circulation in the overworld, and the exchange with the middle-world and troposphere, are controlled by the propogation of Rossby and gravity waves originating in the troposphere, followed by their subsequent breaking, and deposition of momentum in the overworld. This wave drag drives the mass circulation in the middle atmosphere. The radiative forcing in the middle atmosphere actually acts as a spring to resist the dynamical driving. These concepts regarding the relative roles of wave, and radiative forcing, stimulated a number of questions from the audience, because it differs from the intuitive picture of the mean circulation as being a thermally driven circulation. Holton was also questioned about the role of stratospheric vertical motions in controlling tropospheric circulations. He felt that stratospheric circulation is unlikely to have much impact on the tropospheric circulation.

Holton then showed some results relevant to the mean circulation by using 10 years of UKMO data to compute an estimate of the mean vertical upward motion in the tropics (between 20S and 20N) at 100 mbars, and at 200 mbars. These estimates suggest that the stratosphere can be "flushed" by the STE in about 2 years. Also, because the upward mass transport is stronger at 200 mb, than 100 mb, there must be substantial poleward (lateral) mass transport between these two levels. Holton closed with a discussion on the role of convection in controlling water vapor exchange in the tropics, either through convective overshooting directly injecting mass into the stratosphere, or mass transport by the vertical displacement of parcels associated with convectively driven gravity waves. These concepts generated a substantial number of questions and comment (particularly by Dieter Kley) regarding

1) the relative importance of convective injection vs gravity wave transport.

2) The potential for an unrealistic over-estimate of the poleward transport of mass in the lower stratosphere from the method used to estimate the equatorial vertical mass transport.

3) The need to acknowledge the zonal asymettry in the STE. There was no consensus on a resolution of these questions.

G. Kent provided a broad review of aerosol measurements in the stratosphere and upper troposphere. The talk addressed observations of aerosols of both volcanic and anthropogenic origin. The measurements showed variations by two orders of magnitude following major volcanic eruptions. One feature which stimulated some discussion was the residence time (of order 12 months) for these aerosols in the lower stratosphere. After showing the temporal evolution of aerosols distributions following Pinatubo he noted that there have only been a few attempts to model this evolution.

Kent then showed stratospheric column burdens of aerosol and noted the banded structure evident therein, highlighting an equatorial reservoir of aerosol, and accumulation regions at the poles. SAGE II measurements also showed evidence for the convective lofting of tropospheric aerosols, particularly at mid-latitudes in spring and summer. Latitude/height distributions of volcanic aerosol showed little downward transport in equatorial regions. In the extratropics, the upper troposphere showed similar mass loadings to the stratosphere, but with implied residence times in the troposphere of order 2 months. He closed by briefly remarking on the chemical composition of aerosols (sulfur becoming increasingly important higher in the troposphere), and described evidence for the production of ultra-fine aerosols above the ITCZ in the upper troposphere.

The discussion centered upon the evidence for the lifetime for aerosols following their initial injection by volcanoes, where the apparent loss rate of aerosols from the stratosphere may decrease with time. A potential explanation for this effect is that the larger particles fall out first, while smaller particles remain in the stratosphere for longer periods.

S. Borrmann presented new measurements of stratospheric aerosol and polar stratospheric cloud particles obtained from the high altitude research aircraft ER-2. Implications of these measurements for stratospheric dynamics, particle microphysics and heterogeneous chemistry were discussed. It was shown that in a certain time window after the eruption of Mt Pinatubo the aerosols can serve as dynamical tracers with conserved aerosol number and surface area (above 200 K). Later condensation and coagulation were observed to disturb these properties. By comparison of volcanically perturbed and unperturbed airmasses a saturation response was identified in the hydrolysis of N_2O_5, thereby establishing the first direct evidence of this major heterogeneous reaction in the field. The discussion after the presentation centered on the issue of the precise origin and of limitations in the observed linear relationship between ozone and aerosol surface, and on its dynamical consequences.

Finally, J. Lelieveld presented work focussing on ozone in the troposphere. Studies were done in context of the MOGUNTIA model, a three-dimensional model with a comprehensive chemistry, moderate resolution, parameterized convection, and a simple transport scheme based on monthly winds. The ozone distribution was described in context of a budget study, characterizing ozone in terms of its major sources and sinks. One major source is the stratosphere. Nearly equivalent is the net production in the troposphere. However, the individual photochemical production and loss terms are much greater than the sum, both having approximately doubled due to anthropogenic influence. The model shows a considerable increase in tropospheric ozone over the 1850-levels. The discussion concentrated on the role of convection on the NO_x/O_3 chemistry and on the question why the model gives zonally symmetric trace gas distributions in the equatorial regions.

Session 7: Interaction of Aerosols, Clouds and Radiation

Rapporteurs : M. Baker[1] and J. Heintzenberg[2]
[1]University of Washington, Seattle, WA 98195, USA
[2]Institute for Tropospheric Research, D-04303, Leipzig, Germany

The interaction of aerosols, clouds and radiation can influence the climate forcing in the following ways:

- direct radiative forcing by scattering, absorption and emission of radiation by aerosols,
- indirect radiative forcing through changes in cloud properties influencing their radiative properties, and
- hydrological forcing by influencing precipitation development and cloud persistence, including subsequent radiative consequences.

The presentations in this session focused on the first two of these forcings.

R. Charlson presented evidence for the control of the optical depth in the northern hemisphere by anthropogenic sulphate. A succession of 3-dimensional global models was used to quantify the direct radiative forcing of sulphate aerosols and to compare this forcing with that of anthropogenic greenhouse gases. Latest global results are on the order of -0.3 to -0.6 Wm^{-2} for the anthropogenic sulphate versus 2 Wm^{-2} for the greenhouse gases. Although the gases are distributed evenly over the globe, their greenhouse forcing is spatially inhomogeneous for two reasons: (a) regions exist where the atmosphere is already opaque in the relevant bands and (b) the forcing is strongly dependent on surface T. Because of the short lifetimes of aerosol particles and the regional distribution of sources, aerosol forcing is spatially inhomogeneous on regional scales. As a result, the greenhouse gas and aerosol forcings do not simply cancel out. In particular, the anthropogenic effects increase the gradient in forcing between the tropics and subtropics. These findings raise the question: what meteorological processes are sensitive to these gradients in anthropogenically induced radiative forcing? In the discussion it was speculated that the tropics/subtropics gradient in forcing might change the ocean heat transport in the tropics where most of the heat transport is oceanic. As caveats, it was pointed out that is has yet to be shown sulphate is the single strongest contributor to the optical depth and that there are many other aerosol components besides sulphate. Y. Fouquart reminded us of the fact that when aerosols were first introduced into GCM radiation codes the discrepancies among the model predictions went up. He proposed that more measurements are needed of the imaginary part of the refractive index because of its effect on cloud absorption.

Indirect aerosol forcing was investigated in another numerical study reported by A. Slingo translating modeled sulphate mass distributions to aerosol number distributions. Using empirical relations between droplet number concentrations and aerosol numbers authors estimated the indirect radiative forcing of anthropogenic sulphate aerosols from changes in sulphate number before feedbacks took place in the model. In the results, the effect of the anthropogenic aerosols is to give an annual mean decrease in effective radius in low clouds of up to 3 μm and annual mean change in cloud forcing of -1 Wm^{-2}. There was significant indirect forcing in the southern hemisphere because of the higher susceptibility of the clouds there.

Some possible effects of CCN on water vapour budgets were discussed by M. Baker. Attention was drawn to the importance on non-equilibrium effects, cloud physics and chemistry often not considering dynamics and related time scales.

H. Graßl evaluated time series of aerosol parameters and global radiation taken over Germany. Up until 1987 aerosol mass concentrations increased over Hamburg while number concentrations of particle sizes less than 0.2 μm continued to increase. Significant downward trends in global radiation (diffuse + direct solar) over Germany were observed, on the order of 1-2% per decade. From a global ground based aerosol data set maps of aerosol optical depth in the visible were constructed that showed somewhat different distributions than corresponding satellite-derived maps.

With the ECHAM-GCM coupled to a slab ocean model, E. Roeckner performed three equilibrium response experiments (1) removing sulphate, (2) doubling CO_2, and (3) allowing 1.4xCO_2 plus anthropogenic sulphate which is a rough estimate of the increased levels of all greenhouse gases and sulphate aerosol since preindustrial times. Due to the combined radiative forcing in experiment (3), the annual mean warming of the northern hemisphere was significantly smaller (3°C) than that of the southern hemisphere (0.7°C). The sulphate effect was strongest in summer. He calculated a severitivity parameter in $K/W/m^2$ that was smaller in experiment (3) than in the other experiments, possibly because of an asymmetric response of sea-ice and continental snow cover. As to the question of any change in wind speeds in the results it was difficult to see any sign of changes.

Observations to investigate anthropogenic effects on (a) aerosol and cloud microphysical properties, (b) cloud albedo and optical depth were shown by S. Schwartz. He used data from both hemispheres to show the anthropogenic effects on sulphate burdens and optical effects on large spatial scales. He then showed small scale temporal and geographical variability in the observations and suggested that there were clearer anthropogenic signals on these scales. Through both observational and model results he demonstrated the large sensitivity of cloud optical properties to anthropogenic aerosol perturbations. There was a systematic hemispheric difference in effective droplet radii but there was no systematic difference in cloud albedo, possibly for want of appropriate data.

Finally, Th. Peter reported stratospheric observational and numerical results concerning the physical and chemical properties of the particles in the Junge layer. In order to understand discrepancies between measured and calculated ozone reductions post-Pinatubo the author studied the physico-chemical properties of PSC particles which are involved in the ozone chemistry. Using theoretical analyses of concentrated electrolytes and observations of aerosol surface versus temperature he concluded that the particles consisted of H_2SO_4, HNO_3 and liquid water. They also examined balloon-derived particle spectra in stratospheric clouds over the Scandinavian mountain range and deduced that the traditional formation mechanism was inadequate. However, assuming that the cloud particles were liquid droplets gave good agreement with the observation. In the discussion, the question about the importance of chemical versus dynamic factors in controlling cloud formation was raised.

Microphysical and Dynamical Control of Tropospheric Water Vapor

Kerry A. Emanuel
Center for Meteorology and Physical Oceanography
Massachusetts Institute of Technology
Cambridge, MA 02139

Raymond T. Pierrehumbert
Department of Geophysical Sciences
University of Chicago
Chicago, IL 60637

1. Introduction

Water vapor and condensed water are by far the most important greenhouse substances in the Earth's atmosphere, together accounting for the great majority of the net back-radiation from the atmosphere. In addition, clouds reflect incoming solar radiation, constituting about two-thirds of the planet's albedo. The condensed water in clouds is also strongly related to atmospheric water vapor content, occurring only where the actual vapor pressure locally approaches its saturation value. For these reasons, understanding the control of atmospheric water vapor is an essential part of understanding climate change.

Cycling of water through the troposphere takes on the order of 2 weeks, which is much shorter than the time scale of climate variability, so that we may regard the distribution and amount of water vapor and clouds as being in equilibrium with the rest of the climate system. (Residence time of water in the stratosphere is much longer, but stratospheric water has relatively smaller influence on radiative transfer, so we shall not concern ourselves with it here.) We argue here that the water content of the troposphere is controlled by both dynamical and microphysical processes, and that neither of these controls is well understood or represented in climate models. We demonstrate, using simple process models, that climate is sensitive to poorly known microphysical and dynamical processes and argue strongly for a program of research and field measurements designed to help resolve this problem.

NATO ASI Series, Vol. I 35
Clouds, Chemistry and Climate
Edited by Paul J. Crutzen and Veerabhadran Ramanathan

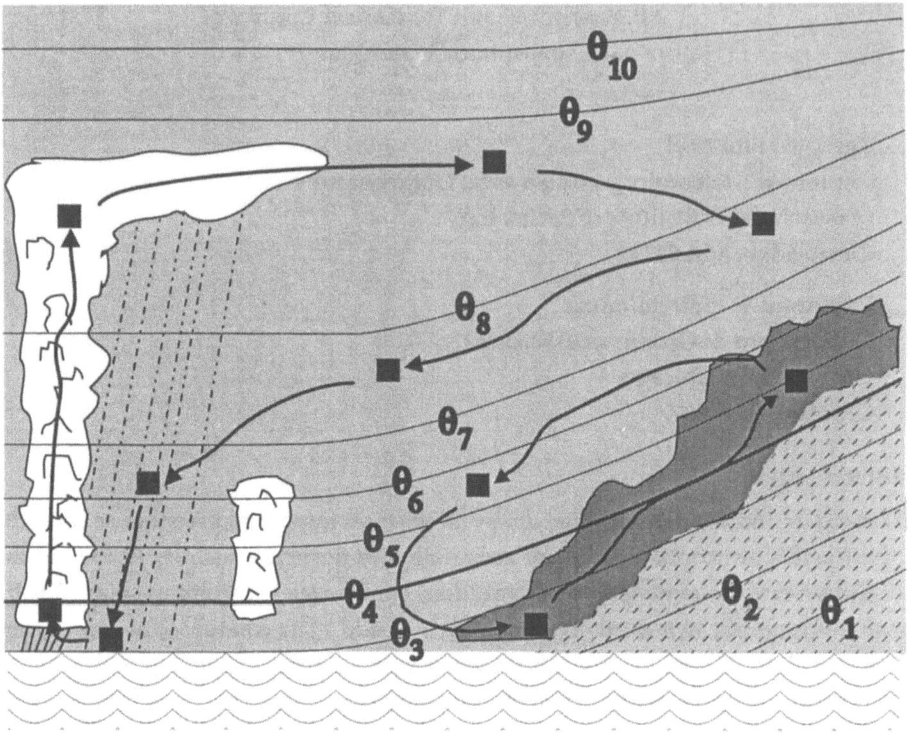

Figure 1. Illustration of Lagrangian trajectories through the atmosphere, showing the importance of microphysical processes in determining the water content of air. The total water content of air flowing out of clouds is set by the fraction of condensed water converted to precipitation, and subsequent moistening in the general subsiding branch is governed by detrainment from shallower clouds and by evaporation of precipitation.

2. Microphysical control

Consider a randomly chosen sample of air from the tropical subcloud layer, which contains the highest moisture content of any air in the atmosphere, and follow it (in a Lagrangian sense) as it moves around in the troposphere. It is advantageous to think of that movement in potential temperature (θ) coordinates, as shown in Figure 1.

Air crosses upward through θ surfaces principally through latent heat release in clouds; mostly in tall cumulonimbus in the tropics and over the oceans in general (even in winter), and in the moist slantwise ascent regions of baroclinic cyclones. Air descends through θ surfaces mostly in rainy downdrafts (at low levels) and through loss of heat by longwave radiation to space.

Now consider the water content of the test sample, averaged over a large ensemble of such samples. As air ascends through θ surfaces, it is saturated and generally contains cloud water. Air that ascends through the depth of the troposphere loses the vast majority of its water content by precipitation; still, it may exit the tops of clouds with condensed water content vastly in excess of the vapor content, owing to the very small saturation vapor pressure of the cold air near the tropopause. As it begins its descent, the sample's water content is the sum of its condensed water and water vapor content when it exits deep clouds.

It is important at this point to recognize that typical ascent velocities in clouds are at least an order of magnitude greater than the magnitudes of descent velocities in clear air, so most of an air sample's lifetime will be spent in its clear descent phase (or in passive stratiform clouds). Thus, much of the problem of determining the atmosphere's water content involves accounting for water outside of *deep* cloud systems.

The descending sample may take many weeks to return to the subcloud layer, during which time it may traverse considerable horizontal distances. As it descends, it is moistened (in a statistical sense) by detrainment from shallower clouds and by the partial evaporation of precipitation falling into it from above. (If these processes were absent, the sample would arrive at the surface with a relative humidity of only a few per cent.)

From this point of view, it is easy to see that the average water vapor content of the atmosphere as a whole is strongly influenced by those cloud microphysical processes that determine how much condensed water air has on detraining from clouds, and by the microphysics of the evaporation of precipitation as it falls through subsaturated air. (It is less easy to see that lateral mixing can strongly redistribute water between samples; this will be the subject of the next section.) Because microphysical processes are poorly understood, and because water vapor content in the upper troposphere is not well measured, there is no reason to believe that these microphysical processes are well represented in climate models. Moreover, a typical climate model has seven or eight levels in the troposphere, vastly inadequate to resolve water vapor variations of four or five orders of magnitude.

Much of the vertical water transport in the tropics is accomplished by deep convection. But the convection schemes in use make extremely crude assumptions about moistening effects of convection, bypassing any attempt at a detailed accounting of microphysical processes, for computational and, perhaps, philosophical reasons. One example is the Kuo scheme (Kuo, 1965, 1974), which simply moistens in proportion to the precipitation rate and the departure of the air from saturation. Another is the Manabe scheme (Manabe et al., 1965) which moistens entire grid boxes to saturation (or near

Table 1. Equilibrium Surface Temperatures, Evaporation Rates, and Values of the Sensible Heat Fluxes Obtained in Experiments With Various Parameters

Run	NL, levels	k_v, $m^2\ s^{-1}$	$C_D v_a$, $m\ s^{-1}$	Δt, s	T, °C	Evp, $mm\ d^{-1}$	F_θ, $W\ m^{-2}$
ESSTD	16	2.0	0.125	900	15.2	3.2	7.4
ES8L	8	...	...	...	20.3	3.4	4.9
ES32L	32	...	...	...	13.6	3.1	9.3
ES2KV	...	4.0	...	...	16.9	3.5	5.6
ES2CD	...	...	0.250	...	15.0	3.8	−2.3
ESDT2	...	...	...	1800	15.3	3.2	7.7

saturation) and precipitates any condensed water without allowing for re-evaporation. The Arakawa-Schubert scheme (Arakawa and Schubert, 1974) in its original form does not allow for re-evaporation of precipitation either. Perhaps the best scheme with some use in climate models is the Betts-Miller scheme (Betts, 1986; Betts and Miller, 1986), which relaxes moisture toward an empirical profile. But there is no physical reason to attach any universality to such a profile.

The sensitivity of one-dimensional, radiative-convective models to the type of convection scheme used, its parameters, and the model numerics has been documented by Rennó et al. (1994). An example of this sensitivity is demonstrated in Table 1, which shows the equilibrium surface temperatures in a one-dimensional radiative-convective equilibrium model in which the surface is a "swamp" (water with no heat capacity). The model uses the radiative transfer scheme of Chou et al. (1991). The scheme uses several broad-band representations of longwave and shortwave radiation, accounting for longwave absorption by water vapor, carbon dioxide, ozone, and clouds and shortwave absorption by ozone and water vapor, also accounting for shortwave reflection by the surface and by clouds. (Cloud-radiation interaction is turned off in these runs.) The convection scheme is that of Emanuel (1991) and is based on an episodic mixing model in which unstable air ascends adiabatically and then mixes with its environment at each model level, forming a spectrum of mixtures each of which then ascends or descends to its level of neutral buoyancy, where it detrains. At each step, a certain fraction of the condensed water is converted to precipitation, which falls and partially re-evaporates, driving an unsaturated downdraft that transports heat and water.

Table 1 shows the sensitivity of the equilibrium surface temperature to NL, the number of model levels, k_v, the vertical diffusion coefficient, $C_D v_a$, the coefficient of surface heat and moisture fluxes, and the model time step. Note that the run with only 8 levels yields a surface temperature 5°C higher than the control run.

Table 2. Equilibrium Surface ("Swamp") Temperatures Obtained in Experiments With Various Microphysical Parameters

	Cloud Microphysical Parameters				
Run	PB_{crit}, mbar	PT_{crit}, mbar	σ_d	σ_s	T, °C
STD	150	500	0.010	0.150	26.0
LGE	...	200	...	...	22.4
LWE	...	700	...	...	29.8
ELE	...	800	...	...	runaway
LSD	...	...	0.100	...	29.8
SSS	...	...	...	0.015	22.9

Table 2 shows the sensitivity of equilibrium surface temperature to microphysical parameters in the model. The most important parameter set is that which governs the conversion of cloud water to precipitation. Provisionally, this parameter is set as a function of the depth of the updraft:

$$\epsilon_i = \begin{cases} 0 & p_{icb} - p_i < PB_{crit} \\ \frac{p_{icb}-p_i-PB_{crit}}{PT_{crit}-PB_{crit}} & PB_{crit} < p_{icb} - p_i < PT_{crit} , \\ 1 & p_{icb} - p_i > PT_{crit} \end{cases} \tag{2.1}$$

where ϵ_i is the fraction of condensed water converted to precipitation at level i, p_i is the pressure at that level, p_{icb} is the pressure at cloud base, and PB_{crit} and PT_{crit} represent the minimum pressure depth below which no water is converted to precipitation, and the maximum pressure depth above which all water is converted to precipitation, respectively.

The other parameters shown in Table 2 include the fractional area covered by precipitating downdrafts, σ_d, and the fraction of precipitation assumed to fall through unsaturated air, σ_s.

Clearly, there is large sensitivity to all of the microphysical parameters, as expected from the arguments presented earlier. Unfortunately, it is extremely difficult to determine the optimal values of such ad hoc parameters, and to evaluate how well the scheme performs in determining water vapor content. The only conceivable way of validating convective schemes for use in climate models is to force them with vertical velocities, surface fluxes and radiation from a sounding array operated for at least 3 or 4 weeks, and to compare the predicted relative humidities to those observed.

Although the above experiments are concerned with cumulus convection, some of the issues that arise clearly carry over to the representation of microphysical processes in extratropical cloud systems. Moistening by such systems must also depend on conversion of cloud water to precipitation and on re-evaporation of precipitation.

The main conclusion to be drawn from these theoretical considerations and one-dimensional experiments is that *there is no reason to believe that any of the extant cumulus parameterizations is performing adequately in its control of atmospheric water vapor.* We shall see that this problem is compounded by horizontal mixing in the atmosphere.

3. Dynamical control

The microphysical processes described above would pose a formidable problem for modeling even if they took place in a relatively smooth larger scale environment. In reality, they are embedded in a dynamical environment of extraordinary intricacy. Advection by spatially smooth but time-dependent velocity fields can easily lead to chaotic wandering of fluid parcels. This generates small-scale structures in the advected field exponentially rapidly and can bring dry and moist air into close proximity in the vertical or horizontal. Nearby air parcels can have radically different Lagrangian histories, which in turn affects their moisture content. In the following, we illustrate some of these processes as manifested in passive tracer mixing calculations carried out on surfaces of constant dry potential temperature θ. A more complete exposition of isentropic mixing properties in the atmosphere can be found in Pierrehumbert and Yang (1993), and some of the effects of condensation are taken up in Yang and Pierrehumbert (1994). Of course, much essential physics is lost when we ignore exchange of fluid amongst θ surfaces. A full understanding of the typical Lagrangian history of air parcels will be achieved only when the ideas presented below are melded with the primarily cross-isentropic mixing processes detailed in the preceding section.

Figure 2 shows the zonal mean disposition of a typical "middleworld" θ surface such as we shall be concerned with, characterized roughly by the range θ between 300K and 380K. Lower, or "underworld" θ surfaces outcrop at the ground instead of in the tropics, and do not generally intersect the tropopause; higher "overworld" surfaces are wholly stratospheric, apart from cross-isentropic processes. The θ surfaces are not as rigidly fixed in space as are isobaric surfaces, but midlatitude baroclinic transients tend to have parcel trajectories that are only somewhat shallower than the slope of the mean isentropes. Hence, it is generally fair to think of the higher latitude portions of the undulating middleworld surfaces as being of consistently higher altitudes than the lower latitude portions.

Formation of *filamentary structures* is the phenomenon of primary interest. The isentropic trajectory problem constitutes a two dimensional, though time-dependent, dynamical system, and is thus characterized by two Lyapunov exponents. These exponents give the exponential rate of growth or decay of the principal axes of the ellipse

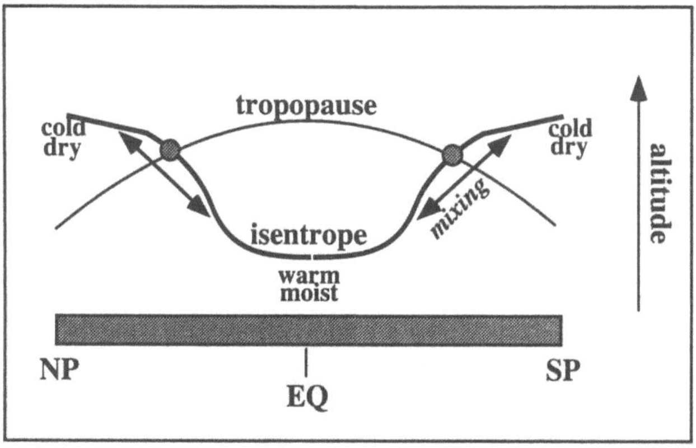

Figure 2. Schematic configuration of a "middleworld" isentropic surface.

into which a small disk of tracers is transformed under action by advection for some specified time. For nondivergent flow, the two exponents must sum to zero, since the area of the disk is preserved. Since the isentropic flow has nonzero divergence, it could in principle lead to situations where both exponents are positive or both negative. Evidently, however, the parcels do not remain in regions of strong divergence or convergence long enough for this to happen. Instead, there is typically one positive and one negative Lyapunov exponent as for the incompressible case, and air parcels are systematically elongated in one direction while they are being systematically contracted in the other. This naturally leads to production of filaments, and to exponential amplification of gradients. Over finite times, the elongation rate is a sensitive function of the initial position of the trajectory, as discussed in Pierrehumbert and Yang (1993).

In Figures 3 and 4 we show how ten days of mixing re-arranges the air on the 315K, 330K, and 350K θ surfaces. The winds used in the calculation are taken from the 12-hourly ECMWF operational analyses for March 1–10, 1993 (the beginning of the CEPEX period). We show snapshots on March 10, with the air colored according to the *latitude* at which it was located on March 1. A great deal of fine-scale structure has been generated after only 10 days. In fact, a direct calculation of the Lyapunov exponents shows that even more fine structure is generated than can be fully resolved in this mixing simulation. The typical midlatitude exponent of .7/day implies that a 500-km-wide smooth feature would be elongated into a filament 500,000 km long (which would have to be folded many times to fit on the globe) and only 500 m thick after 10 days. The explicitly resolved filaments in Figures 3 and 4 arise from trajectories with anomalously weak stretching.

The joint action of planetary waves, shear in the jet streams, and especially synoptic eddies causes rapid and efficient mixing within the two extratropical "ergodic"

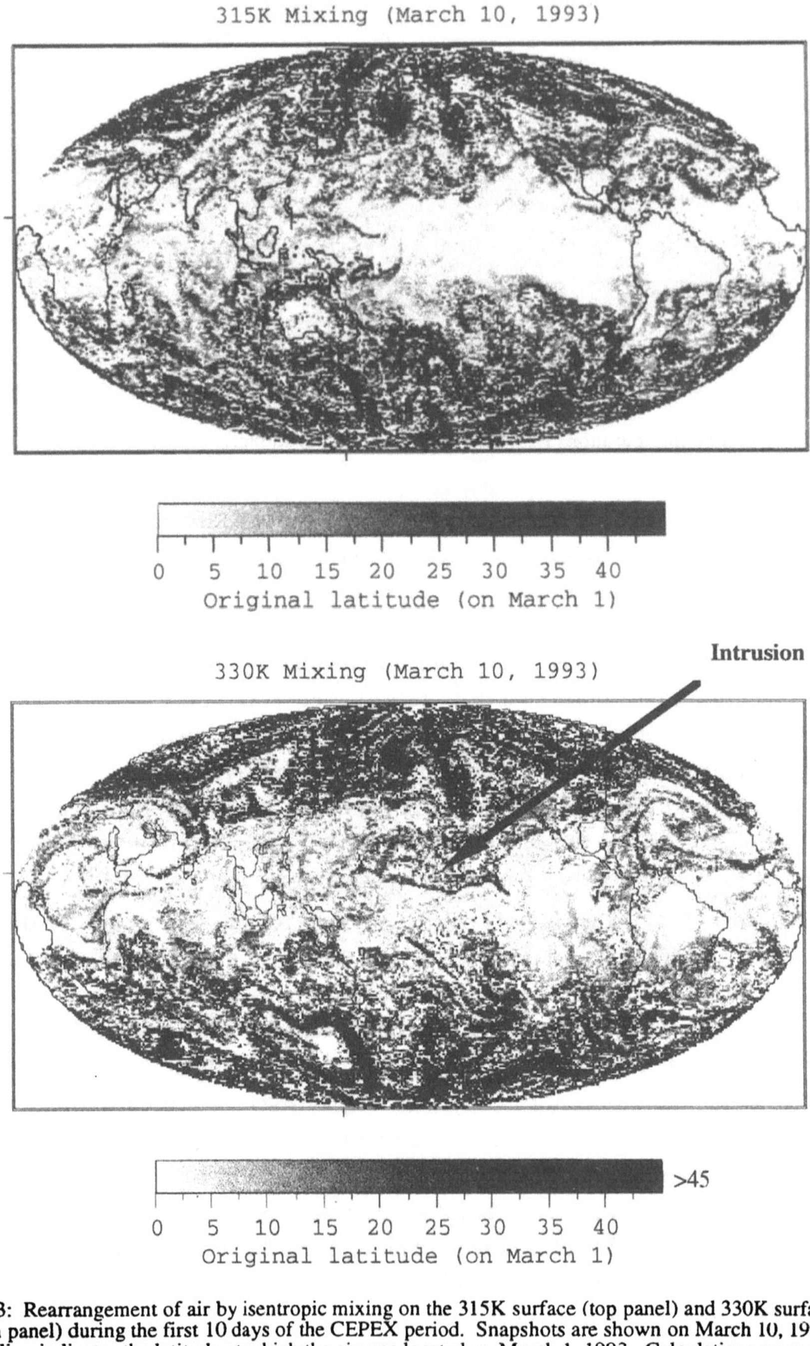

Figure 3: Rearrangement of air by isentropic mixing on the 315K surface (top panel) and 330K surface (bottom panel) during the first 10 days of the CEPEX period. Snapshots are shown on March 10, 1993, and the shading indicates the latitude at which the air was located on March 1, 1993. Calculations were carried out using ECMWF operational wind analyses interpolated to isentropic surfaces.

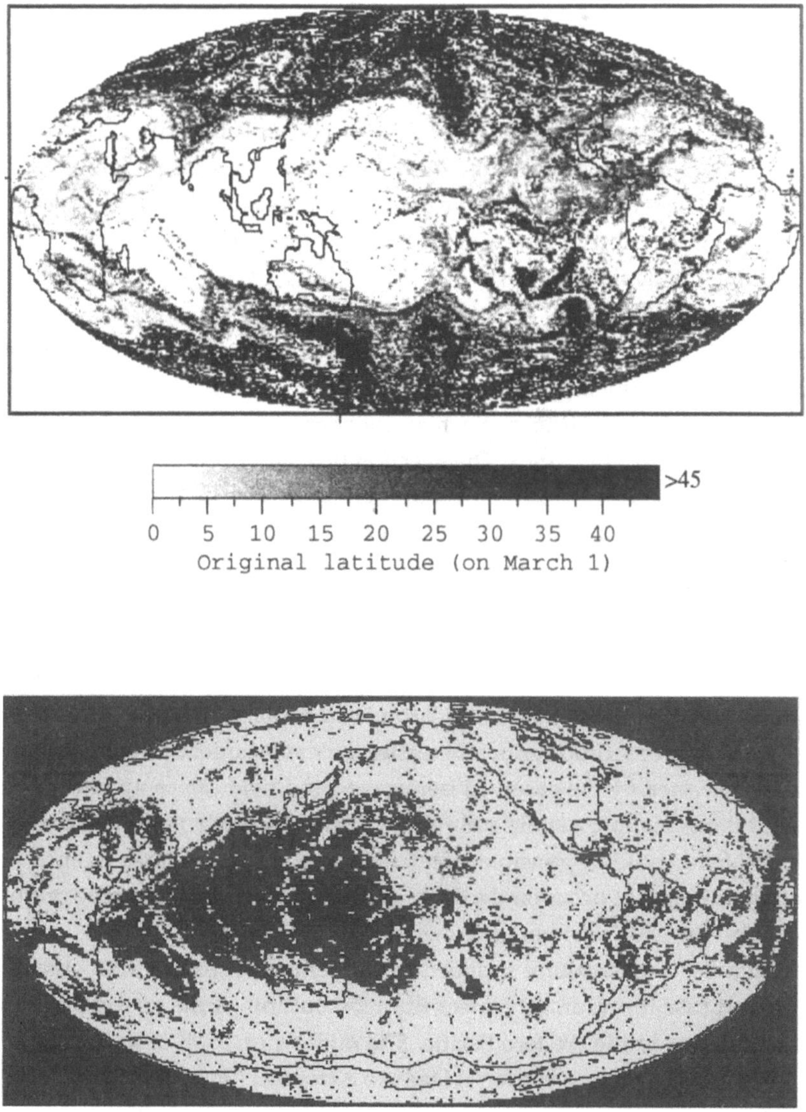

Figure 4: (Top panel) As for Figure 3, but for the 350K surface. The lower panel shows the region (shaded in black) which has been swept clear of particles during the first 10 days.

or "mixing" zones. The meridional displacement of air parcels in conjunction with the upward slope of the isentropes systematically processes air through the high-latitude, cold portions of the isentropic surface (Kelly, Tuck, and Davies, 1991; Yang and Pierrehumbert, 1994), and constitutes a potent drying mechanism. Dry air produced in this manner is distributed throughout the midlatitudes in the form of filaments. General circulation models which fail to resolve these filaments would mix the dry air into adjoining moist air, spuriously lowering the relative humidity of the moist filaments and thus inhibiting condensation.

The extratropical mixing zone is bounded by a subtropical barrier, across which mixing is much weaker. Nonetheless, vigorous recirculating eddies of the subtropical "surf zone" episodically inject dry subtropical air and dry air of high-altitude midlatitude provenance into the tropics. Such an event can be seen in the Central Pacific on the 330K level shown in Figure 3. The injection process is very intermittent in the vertical, and (in the Central Pacific) shows up neither on the 350K nor 315K surfaces. It should thus leave its mark on suitable Central Pacific soundings in the form of a distinct midtropospheric layer of anomalously dry air. Aside from their direct effect on tropical humidity, the dry air incursions affect the relative humidity of the air through which precipitation is falling, and hence influence the precipitation efficiency.

Our mixing calculations were performed using a fully Lagrangian method, in which particles tagged with their initial latitude were placed initially on a 1-degree longitude by 1/2 degree latitude grid, whereafter their subsequent positions were tracked as they are advected by the observed isentropic wind field. Since no new particles are added in this calculation, a sufficiently strong and sufficiently persistent divergence of the velocity field could in principle sweep an entire large-scale area clear of particles. This situation does not occur on the 315K or 330K surface, but it is very prominent in the tropics on the 350K surface, as shown in the lower panel of Figure 4. Persistent divergence implies a mass source supplied to the isentropic surface from other surfaces; in the present case it presumably arises from the outflow of the upward branch of the Walker circulation in the Western Pacific. The outflow on 350K is so strong that it creates a novel type of mixing barrier; it leads to a tropical "forbidden zone" into which alien air cannot readily penetrate. Aircraft or sonde observations in this region should show a fairly uniform body of air of low-altitude provenance. Longer-term integrations (not shown) indicate that the air mass remains quite isolated until March 20, whereafter appreciable amounts of air from elsewhere in the tropics begin to leak in.

Finally, there is the matter of mixing *within* the tropics due to large-scale tropical transients. One is accustomed to the simple Eulerian picture of the Hadley or Walker circulation in which air goes up, outwards and down, ultimately closing the loop in a low-level return current. This would lead to extremely dry air in the subsiding branches, in the absence of compensating moistening processes. However, trajectories based on time-averaged winds gives a misleading picture of the true Lagrangian histories of air parcels. Tropical transients are weak compared to extratropical transients, but the mean flow in which they are embedded is also weak, and so considerable mixing is still possible, albeit on a slower time scale than prevailing in the extratropics. The signature of lateral mixing is clear in the tracer calculations shown in Figures 3 and 4. Lateral mixing of this type may export moist air from the warm-pool atmosphere, and thus provide an important moisture source for the regions of time-mean subsidence in the Hadley and Walker circulation. This mechanism provides an alternative to the proposal of Sun and Lindzen (1993), who imply that evaporation of hydrometeors is the primary moistening mechanism throughout the tropics. Their proposal seems problematic, because it is far from clear that there are enough deep precipitating systems in the vicinity of the large-scale subsiding regions to provide the necessary moisture. While evaporation of hydrometeors is undoubtedly important in moistening air in the vicinity of deep cumulonimbus clouds, as we saw in the previous section, lateral mixing is probably important for redistributing this moist air throughout the descent regions. If lateral intratropical mixing indeed turns out to be a significant moisture source, an additional burden would be imposed on the climate modeling enterprise: climate models would need to accurately reproduce tropical transients as well as the time-mean Hadley and Walker circulations. The representation of such features is not wholly satisfactory even in operational forecast models, and is still less so at the lower resolutions currently typical of climate-oriented GCM's.

4. Summary

We argue here that the distribution of water vapor in the atmosphere is controlled by cloud microphysical processes and by detailed advective processes which tend over time to lead to fine-scale filamentary structures. There is little if any reason to believe that microphysical processes are handled in any adequate way within either parameterized or explicit clouds within general circulation models, or that the eddy activity in such models is correctly advecting water vapor, particularly in the tropics. Since water vapor and the cloudiness which is quintessentially related to it are the principal greenhouse substances in the atmosphere, it is clear that a far better understanding of observed water substance distributions will be necessary before GCM's can be tested in a meaningful way. We believe that it is critical that suitable field programs be designed to quantify the distribution and budget of atmospheric water substance, particularly in the upper troposphere.

5. References

Arakawa, A., and W. H. Schubert, 1974: Interaction of a cumulus cloud ensemble with the large-scale environment, Part I. *J. Atmos. Sci.*, **31**, 674–701.

Betts, A. K., 1986: A new convective adjustment scheme. Part I: Observational and theoretical basis. *Quart. J. Roy. Meteor. Soc.*, **112**, 677–691.

Betts, A. K., and M. J. Miller, 1986: A new convective adjustment scheme. Part II: Single column tests using GATE wave, BOMEX, ATEX and arctic air-mass data sets. *Quart. J. Roy. Meteor. Soc.*, **112**, 693–709.

Chou, M.-D., D. P. Krats, and W. Ridgway. 1991: Infrared radiation parameterization in numerical climate models. *J. of Climate*, **4**, 424–437.

Emanuel, K. A., 1991: A scheme for representing cumulus convection in large-scale models. *J. Atmos. Sci.*, **48**, 2313–2335.

Kelly, K. K., A. F. Tuck, and T. Davies, 1991: Wintertime asymmetry of upper tropospheric water between the Northern and Southern Hemispheres. *Nature*, **353**, 244–247.

Kuo, H.-L., 1965: On formation and intensification of tropical cyclones through latent heat release by cumulus convection. *J. Atmos. Sci.*, **22**, 40–63.

Kuo, H.-L., 1974: Further studies of the parameterization of the influence of cumulus convection on large-scale flow. *J. Atmos. Sci.*, **31**, 1232–1240.

Manabe, S., J. Smagorinsky, and R. F. Strickler, 1965: Simulated climatology of a general circulation model with a hydrologic cycle. *Mon. Wea. Rev.*, **93**, 769–798.

Pierrehumbert, R. T., and H. Yang, 1993: Global chaotic mixing on isentropic surfaces. *J. Atmos. Sci.*, **50**, 2462–2480 .

Rennó, N. O., Emanuel, K. A., and P. H. Stone, 1994: Radiative-convective model with an explicit hydrological cycle, Part I: Formulation and sensitivity to model parameters. *J. Geophys. Res.*, **99**, 14429–14441.

Sun, D.-Z., and R. S. Lindzen, 1993: Distribution of tropical tropospheric water vapor. *J. Atmos. Sci.*, **50**, 1643–1660.

Yang, H., and R. T. Pierrehumbert, 1994: Production of dry air by isentropic mixing. *J. Atmos. Sci.*, (in press).

Interactions of Radiation and Microphysics in Cirrus

J. Heintzenberg[1], Y. Fouquart[2], A. Heymsfield[3], J. Ström[4] and G. Brogniez[2]

[1] Institute for Tropospheric Research, Permoserstr. 15, D-04304 Leipzig
[2] Laboratoire Optique Atmosphérique, URA CNRS 713, Université de Lille 1, F-59655 Villeneuve d'Ascq France
[3] National Center for Atmospheric Research, MMM Division, Boulder, CO 80307-3000, USA
[4] Department of Meteorology, Stockholm University, S-10691 Stockholm, Sweden

1. INTRODUCTION

Cirrus clouds are rapidly gaining increased attention as an important component of the climate system. There are some solid physical reasons for that: cirrus are extensive in space and time. In addition, they are cold enough to cause a strong greenhouse effect. However, the problem of quantifying their influence on the present climate has remained essentially unresolved and we are even farther from being able to predict their response to a changing climate.

While the issue of understanding cloud feedback mechanisms is not limited to cirrus these clouds have some characteristics which complicate the problem. Obviously they are too high for easy access to in-situ observations. More important here is that they are composed of ice crystals of various habits (bullets, columns, plates) which cause optical phenomena (halos, arcs, parhelions, anthelions...) which are evident in most observations of cirrus. Three, at the low temperatures at cirrus altitudes, only little water vapour is available for cloud formation and thus for experimental detection. These characteristics will influence the radiative properties of cirrus cloud layers and thus their life cycle and their connection with the climate system.

NATO ASI Series, Vol. I 35
Clouds, Chemistry and Climate
Edited by Paul J. Crutzen and Veerabhadran Ramanathan

2. MICROPHYSICAL PROPERTIES OF CIRRUS CLOUDS

2.1. Cirrus formation processes

In the altitude region where cirrus formation usually takes place we have an aerosol ranging in the number concentration between 10^3 and 10^6 L^{-1}. According to what we know about these particles, they mainly consist of sulphate, (by number). Due to the limited supply of appropriate neutralising gases e.g., NH_3, the sulphate should be present in the form of sulphuric acid or as ammonium bisulphate. In the upper troposphere a larger fraction of the oxidation of trace gases can be expected to take place in the gas phase as compared to the lower troposphere where much of the secondary aerosol production takes place in cloud water. Consequently, we expect gas phase oxidation products in the cirrus region which are able to condense on non-hygroscopic particles, thus converting even this aerosol fraction to a state which facilitates its participation in cloud formation processes.

For example, Detwiler (1989) noted that this scenario provides ample possibilities for the observed concentrations of cirrus crystals, ($<10^4$ L^{-1}, cf. section 2.2) to form through a chain of processes leading via liquid phase condensation - dilution through condensational growth - and freezing with subsequent sublimation to visually observable cirrus crystals.

The importance of this formation process for the cirrus region has yet to be demonstrated. Indirect evidence exists though, for cirrus going through a freezing haze state before reaching crystal sizes and habits that cause the display of its prominent optical features. With the advance of crystal sizing techniques, the observed number concentrations keep increasing towards smaller sizes. Below $\approx 50\mu m$ in linear dimension, particles approach simpler, more spherical shapes with decreasing size. The "sub visual" cirrus that has been discussed for some time, (Platt, et al., 1987; Sassen and Cho, 1992; Sassen, et al., 1989) could be a sign of the haze transition in newly forming and evaporating cirrus.

Cirrus shields forming as the result of deep convective clouds will go through warm or mixed cloud phases before reaching the cirrus level. Thus, their formation process will differ in the sense of a larger likelihood of an involvement of special ice nucleating particles. On the other hand, on a global scale we do not expect this type of cirrus formation to control total cirrus cloudiness.

2.2 Integral cirrus properties

The integrals of the first three moments of the crystal size distribution are the most basic microphysical cirrus properties. Here we focus on moment zero yielding total number (N_c), an average crystal length (L_c), derived from moment one and condensed water content (CWC) derived from the third moment of the distribution. However, already at this level our knowledge is rather limited. Most data have been derived from instruments which do not sense particles with dimensions less than 30 μm while crystal size distributions tend to show increasing number concentrations towards that lower size limit. Most CWC data, on the other hand, are derived from shadow measurements of the same type of sensors by making varying assumptions about particle shapes. In a recent cirrus summary, Dowling and Radke (1990) quoted values of 10^{-4} - 10^{4} L^{-1} for N_c, $2x10^{0}$ - $8x10^{3}$ μm for L_c and 10^{-1} - $1.2x10^{3}$ mgm^{-3} for CWC. Besides meteorological factors like cloud height and temperature, technical reasons are responsible for the substantial variability in the reported integral properties, e.g., the influence of crystal habit on the optical shadow measurements.

Recently, several authors (Noone, et al., 1993; Ström and Heintzenberg, 1994 and Ström, et al., 1994) reported CWC data sets that were not dependent on assumptions about the crystal shape. This approach uses the counterflow virtual impactor (CVI) technique to sample cloud elements irrespective of their phase and shape, (Ogren, et al., 1985). By counting the crystal residues with condensation particle counters N_c is also measured and diameters of mean crystal mass (DMM) are calculated by dividing CWC by N_c. These new measurements cover the aerodynamic size range of five to ≈50μm. Flight averages of their data are collected in Table 2.1. In recent flights over Central Europe, even higher crystal concentrations (≈ 5000 L^{-1} with CWC ≈ 10 mg m^{-3}) were encountered in 1994 with the CVI technique in cold cirrus.

Table 2.1 Flight averages of integral cirrus properties derived with a counterflow virtual impactor. N_c = crystal concentration as measured by their residuals after evaporating the water of the cirrus crystals, DR = diameter of the residual particles, CWC = condensed water content, DMM = diameter of mean mass.

Experiment	N_c L^{-1}	CWC mg m^{-3}	DMM µm
ICE-89	15	1.3	50
	24	2.5	54
	20	1.9	48
DR > 0.14µm	11	1.1	50
EUCREX-92	130	1.6	28
DR > 0.015µm	380	3.7	13

2.3 Size distribution of cirrus particles

Cirrus size distributions follow a trend with temperature that is indicative of the decrease in amount of water vapor available for condensation with decreasing temperature. As the temperature decreases, the mean particle size decreases. The size distribution can be represented by exponential or power-law functions with negative slope, i.e. the concentration decreases with increasing diameter D. It has been found that the slope of the size distributions becomes steeper with decreasing temperature. For example, Heymsfield and Platt (1984) characterised the size distributions by an equation of the form $N = A\ D^{-B}$, where A and B, the intercept and spectral slope are found from curves which have been fit to experimental size distribution data. The coefficient B is shown as a function of temperature in Table 2.2.

The largest particle sizes in a given size distribution increase with increasing temperature. When temperatures exceed -40°C, the largest particles in the distributions often are larger than 1 mm, while below -60°C they are at most 10 to 100 µm.

Table 2.2 Slope of the number size distribution of two different types of cirrus crystals as a function of temperature.

Temperature range (°C)	Slope of the size distribution	
	in situ produced cirrus	Tropical cirrus (CEPEX)
-20 TO -30	-2.6	-2.7
-30 TO -40	-2.2	-3.0
-40 TO -50	-3.2	-3.5
-50 TO -60	-3.8	-3.5

The measurements also indicate that cirrus crystal number concentrations fall in the range 10^1 to 10^3 L^{-1}, most frequently from 10^1 to 10^2 L^{-1}. In general, the concentrations decrease with decreasing temperature. However, the magnitude of the total concentrations of the size distributions are uncertain because of the inability to measure reliably ice crystal size distributions below about 30 μm in length. There is evidence of high concentrations of ice crystals with lengths less than 30 μm, especially at temperatures below about -45°C, or when particles form in an ice free region, (cf. section 2.2).

The information reported above is mostly from cirrus formed in-situ. Comparatively little is known about the properties of the size distributions in cirrus anvils, in part because of the difficulty in reaching the cloud tops and in part because of instrument limitations. Several tropical cirrus cloud studies show that the properties of the size distributions behave much like the properties of particles in cirrus produced in-situ. However, as tropical cirrus tops can reach much lower temperatures than have been found for cirrus produced in-situ, mean particle sizes have been found to be smaller. For example, a study of a cirrus layer at -80°C by Heymsfield (1986) using a Formvar replicator to measure the concentration of small ice crystals showed that mean particle sizes were ten micrometer and below and maximum particle sizes were below 50 μm. Concentrations were on the order of 10^1 L^{-1}. Recently, Knollenberg, et al. (1993) reported on observations of ice crystals in anvils associated with tropical convection at temperatures from -60 to -90°C. Newly developed instruments which could measure down to 0.1μm in size were used. Near anvil tops, the measurements revealed number densities typically greater than 10^4 L^{-1}

and as high as 10^5 L^{-1}. The slope of the size distribution in anvils appears to behave in a manner similar to that found for cirrus produced in-situ. A first examination of the CEPEX data by Dr. McFarquhar of Scripps reveals that the slopes are similar (cf. Table 1), but his work shows that the distributions may not follow a power law e.g., they may rather be log-normal.

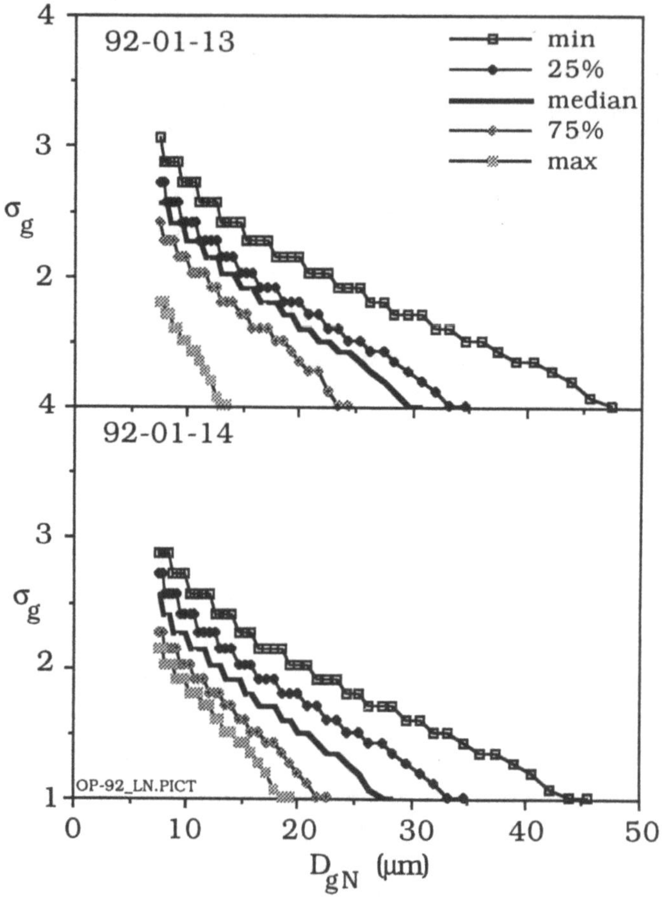

Fig. 2.1 Flight Averages of crystal number size distributions inverted from N_C and CWC during the EUCREX[1] -92 experiment

A log-normal approximation of the crystal size distribution has also been made with the EUCREX-92 CVI data. With the help of the relationships

---

[1] EUCREX = European Cloud and Radiation Experiment

given in Heintzenberg (1994), the parameters geometric mean diameter D_{gN} and geometric standard deviation s_g can be determined from the measured integral values of the 0th (N_C) and 3rd moment (CWC) of the size distribution. For the EUCREX-92 data in Table 2.1 the corresponding log-normal parameters are displayed in Fig. 2.1. To show the range of variability, the distribution parameters have been inverted for the extremes and for the quartiles of the two flights

2.4 Height- and temperature dependence of the microphysical properties

Aircraft in-situ measurements of cloud microphysical properties in tropical cirrus anvils from the Central Equatorial Pacific Experiment (CEPEX) were used to characterise the dependence of the microphysical properties on temperature. The eventual goal of this research is to understand further the connection between cloud microphysical properties and cirrus albedo.

A cirrus anvil is an upper-level ice cloud produced by detrainment of cloud material from a region of deep convection. Data from vertical profiles through three anvil clouds and from many hours of penetrations through other cirrus clouds were processed, analysed, and stored in a large data file. Height and temperature tendencies were then evaluated.

In general, particle size is found to decrease with decreasing temperature. This tendency can be seen from a parameter relevant to cloud albedo, the "effective diameter", D_e, which is twice the ratio of the particle volume to the particle cross-sectional area (cf. Fig. 2.2). In the mean, D_e decreases by about an order of magnitude over the temperature range -20 to -70°C. The wide scatter in the data set can be attributed to variability in cloud dynamics and time from transport of the cloud elements into the anvil.

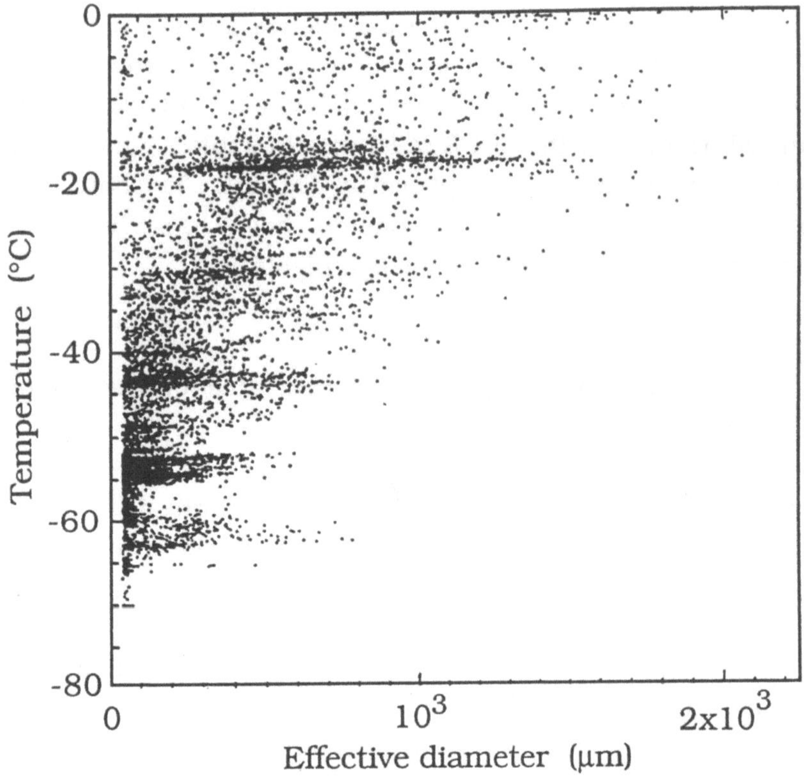

Fig. 2.2 Effective particle diameter (ratio of twice effective radius, which is the ratio of particle volume to area) versus temperature for all 6 second (about 1 km) data points from CEPEX.

Ice crystal habits show considerable variability. The particles are sometimes found to be dense, small ice spheres, sometimes columns, plates, and spatial crystals, with sublimation evident at times as judged from smooth crystal edges (cf. Fig. 2.3).

Condensed water content (CWC) also decreases in the mean, by over an order of magnitude as the temperature decreases from about -35 to -65°C (cf. Fig. 2.4). The CWC's are estimated from particle size distributions measured by Particle Measuring Systems' (PMS) 2D probes (Knollenberg, 1976). Considerable variability in CWC spanning four orders of magnitude is observed at all temperatures. The mean condensed water content within a distance of about 500 m from anvil top ranges from 10^{-4} to 10^{0} g m^{-3} with a mean value of about 10^{-2} g m^{-3}. In the vertical direction, condensed water content shows an increase from cloud top to mid-cloud levels, followed by a decrease below (cf. Fig. 2.5).

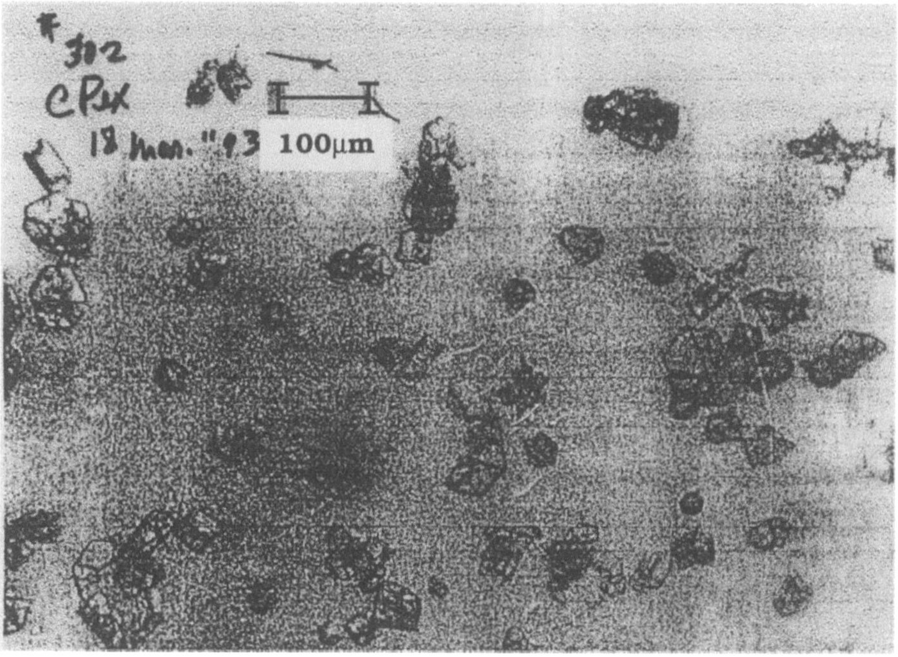

Fig. 2.3 Microphotograph of ice crystals collected in oil coated slides on 16 March 1993.

We have placed an upper limit on the underestimate of the CWC in sizes below the 25µm sizing threshold of the PMS probes by assuming that all size distributions measured by the forward scattering spectrometer probe (FSSP, sizing from 0.4 to 21 µm) are accurate. However, data from this probe are known to be overestimates of true concentrations when in the presence of ice (Gardiner and Hallett, 1985). The analysis shows that we might have appreciably underestimated total CWC's, by more than 50%, when 2D probe CWC's fall below 10^{-3} g m^{-3} but the underestimate is at most 20%, when 2D CWC's were above 10^{-2} g m^{-3}.

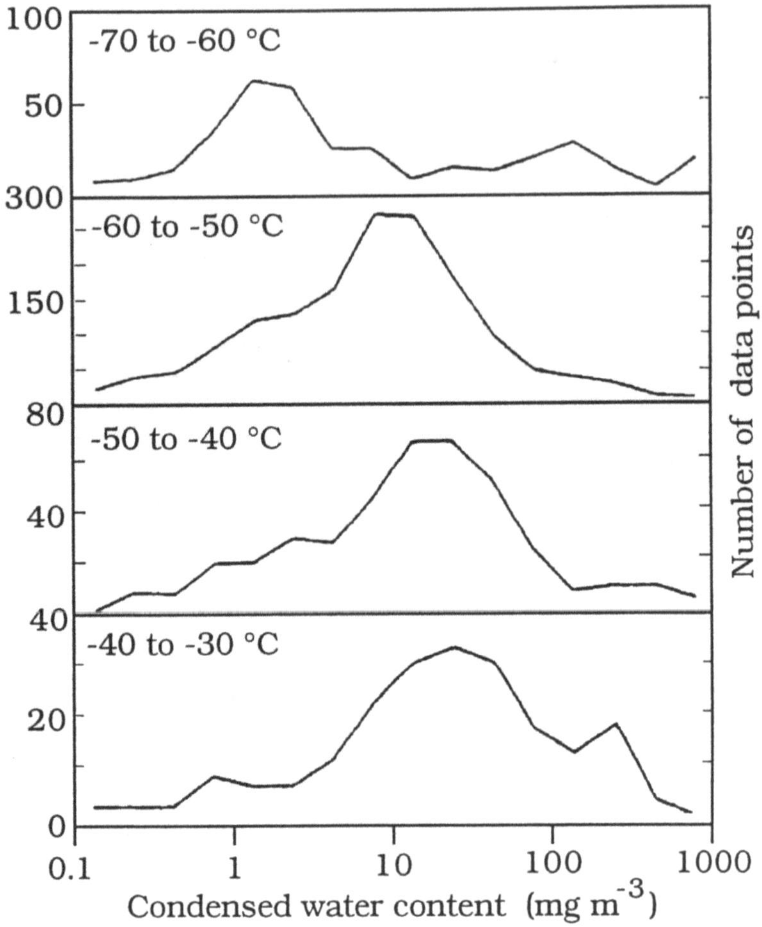

Fig. 2.4 Distribution of condensed water content by temperature interval. The ordinate gives the number of 6 second data points producing a given ice water content within the temperature interval.

Particle cross-sectional areas per volume are measured directly by the 2D probes, but underestimates might have been significant in regions of low optical depth, when small ice crystals might have been present in appreciable concentrations. Possible underestimates were evaluated by combining the FSSP and 2D data. This analysis shows that the cross-sectional area might have been significant when optical depths fall below about 0.1.

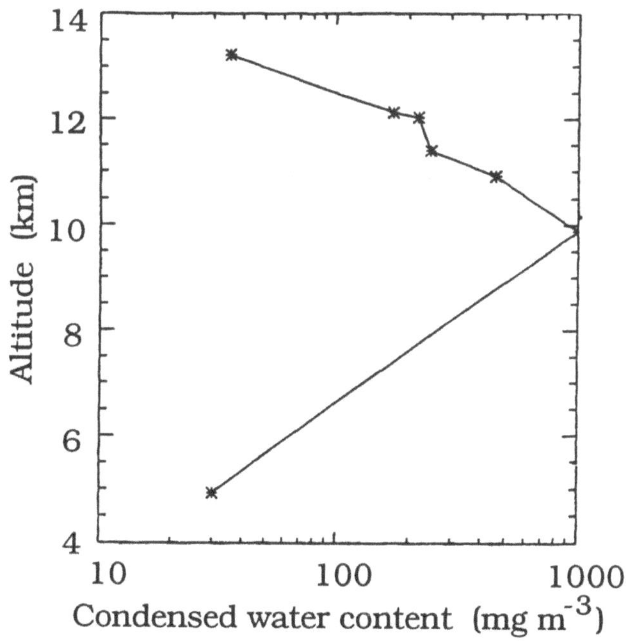

Fig. 2.5 Plot of penetration-averaged condensed water content per penetration versus height on April 4th 1993. Measurements were made just the Northeast of the northern main island of Fiji.

From Flights on ten days, with 9.2 hours within 1/2 km of anvil top, microphysical measurements show:

1. A high degree of variability in all parameters.
2. A decrease in particle size with decreasing temperature (height).
3. CWC at anvil top: 10^{-4} to 1 g m^{-3}, with a mean of 0.01 g m^{-3}.
4. CWC significantly underestimated (>50%) from 2D probes below 10^{-3} g m^{-3}.
5. Particle cross-sectional area significantly underestimated when optical depth through 1 km is 0.1.
6. Crystal habits: Dense, spherical (frozen drops), columns, plates, spatial crystals, some sublimation evident.

3. RADIATIVE TRANSFER IN ICE CRYSTALS

We are interested in cirrus optical properties from two points of view: (1) the radiative transfer through cirrus and (2) the analysis of remote observations of cirrus. In this section, we shall review the state of knowledge of cirrus optical properties emphasising the consequences of the non-sphericity of cirrus particles. The main questions which we address here are:

1. To which extent do cirrus optical properties differ from those of liquid water clouds?
2. What are the possibilities and limitations of the methods presently available to calculate the transfer of radiation through non spherical crystals?
3. What are the simple approximations or the parameterizations which are presently available to treat radiative transfer in cirrus clouds for routine calculations in climate modelling or remote sensing analysis?

3.1. Specific optical properties of cirrus

3.1.1. Transmission and reflection

At the low temperatures which are typical of cirrus, the mixing ratio of water is always small. Thus, although their vertical extension can reach several kilometres, cirrus clouds are most often optically thin. Additionally, and as a consequence of the small ice water contents, cirrus are horizontally inhomogeneous at scales of one kilometer or less. These properties make cirrus clouds quite often semi-transparent on satellite images. As a consequence, most cirrus clouds are undetectable in the visible channels of satellite sensors and difficult to distinguish from lower (and warmer) clouds on thermal infrared images. On the other hand, this semi transparency makes them easy to identify through differences in brightness temperature.

The physical basis of this approach is simple: Molecular absorption by water vapour and other trace gases is wavelength dependent so that the transmission of the clear atmosphere is different in channel four (10.5 -

11.5 μm) and channel five (11.5 - 12.5 μm) of the Advanced High Resolution Radiometer (AVHRR) on the NOAA satellites. The same holds for channel three (3.7 μm). Therefore, the differences between channels four and five accumulate in the right hand corner of a scatter plot that shows them as a function of brightness temperature (cf. Fig. 3.1).

Most lower clouds are optically thick and are opaque in the infrared. Thus, the atmospheric column as seen from satellite will not reach below the cloud tops. In this case, the brightness temperature difference is smaller than in clear atmospheres. It is even zero for high opaque clouds, thus explaining the data points in the lower left hand corner of Fig. 3.1

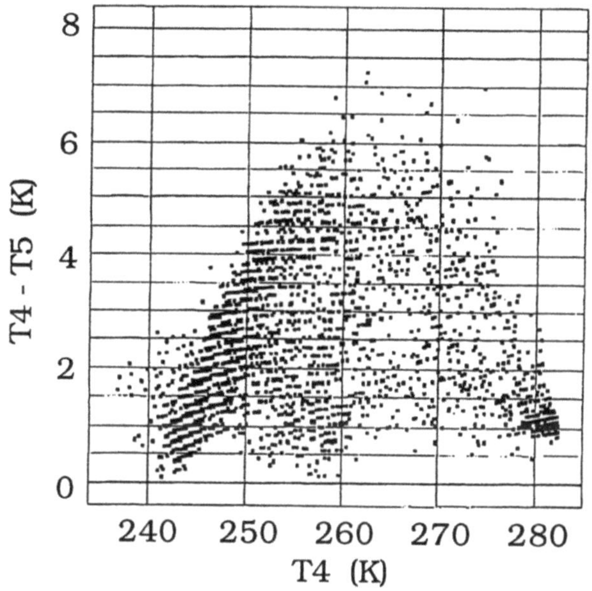

Fig. 3.1 Scatter plot of differences in brightness temperatures in Channels four and five as a function of brightness temperature in channel 4 of AVHRR in the NOAA-11 satellite image for the ICE experiment on October, 18, 1989,

The emissivities of semi-transparent cirrus clouds are spectrally dependent. Thus, the emissivities of the underlying atmosphere is modulated differently by the cloud in both channels, yielding the intermediate results in Fig. 3.1.

Assuming that all pixels have the same microphysical cloud characteristics and that only the total condensed water path varies from pixel to pixel, some authors attempted to derive the size of cirrus particles from the shape of the histogram of brightness temperatures. Inoue (1985) was the first to analyse histograms of brightness temperature differences but he considered purely absorbing clouds. Wu (1987) showed that the differences in brightness temperature are sensitive to the crystal size distributions whereas Parol et al. (1991) considered both scattering by spheres and cylinders and found that neglecting scattering led to systematic underestimates of particles sizes. Lin and Coakley Jr (1993) also used differences in brightness temperature in their cloud retrieval algorithm.

The spectral variation of the imaginary part of the refractive index of ice differs significantly from that of liquid water in the near infrared. In this spectral region, ice clouds are more absorbing than liquid water clouds. This difference has to be taken into account when considering the radiative transfer in cirrus in modelling of cloud life cycles as well as in calculations of vertical profiles of heating rates in the presence of cirrus.

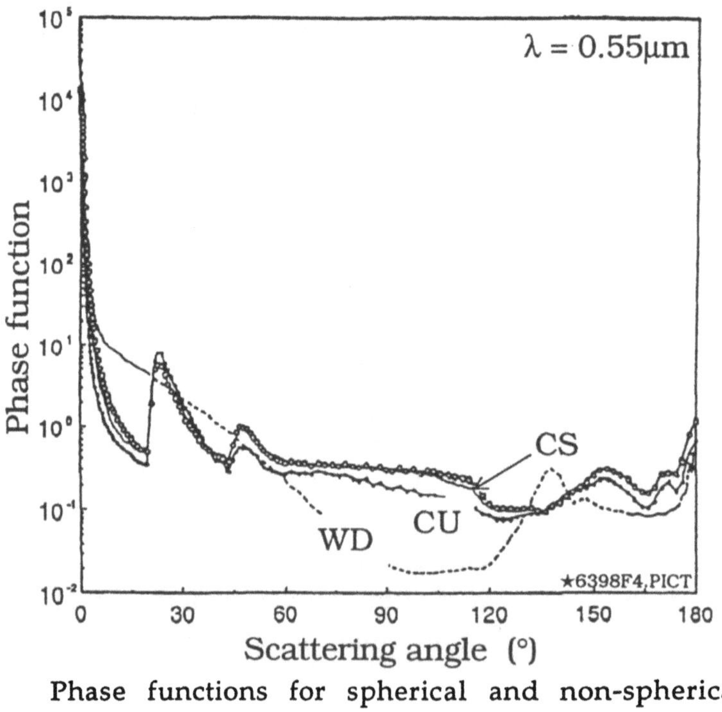

Fig. 3.2 Phase functions for spherical and non-spherical cloud particles, adapted from Minnis et al. (1993). WD = water droplets; CU = cirrus uncinus; CS = cirrostratus.

The non sphericity of ice particles results in various optical phenomena which are typical of cirrus clouds. In some cases these phenomena may have significant effect on either remote sensing analysis or/and on the radiative impact of these clouds. There is a large variety of shapes of ice crystals. Krupp (1991) presented results of observations of ice crystals with a holographic technique during the ICE-89 experiment. These observations were performed on the C-130 research aircraft of the British Meteorological Research Flight Division. In addition to identifying plates, columns, bullets and multi-column particles, these observations confirmed that the crystal size depends on temperature.

3.1.2 The phase function

Haloes, various arcs, parhelions and other optical phenomena are characteristics of ice clouds. More precisely, they are typical of extended cirrus decks since a large spatial extension is necessary for being able to observe these phenomena. They indicate the presence of plates and this particularity is obviously present in the scattering phase function of hexagonal ice crystals, (cf. Fig. 3.2). Note that ice clouds exhibit much more lateral scattering than water clouds but neither rainbow nor glory except for a slight maximum near 155°. The peaks in the forward lobe correspond to the halos at 22° and 46°. For ice crystals the forward peak is much larger than for spheres due to (1) larger cross sections and (2) direct transmission through the parallel faces of the plates.

The relative intensity of the halo gives some information on crystals size. Fig. 3.3, taken from Brogniez, et al. (1994) shows a comparison of measured and calculated scattering phase functions. The measurements were made during the ICE-89 experiment with a forward scattering photometer. This instrument is simply a solar radiometer with a very narrow field of view. By scanning the sky at constant elevation it measures a part of the angular distribution of scattering. If multiple scattering can be neglected, the data can be evaluated to derive the phase function. In general, this is not the case and corrections must be applied in an iterative process to account for multiple scattering.

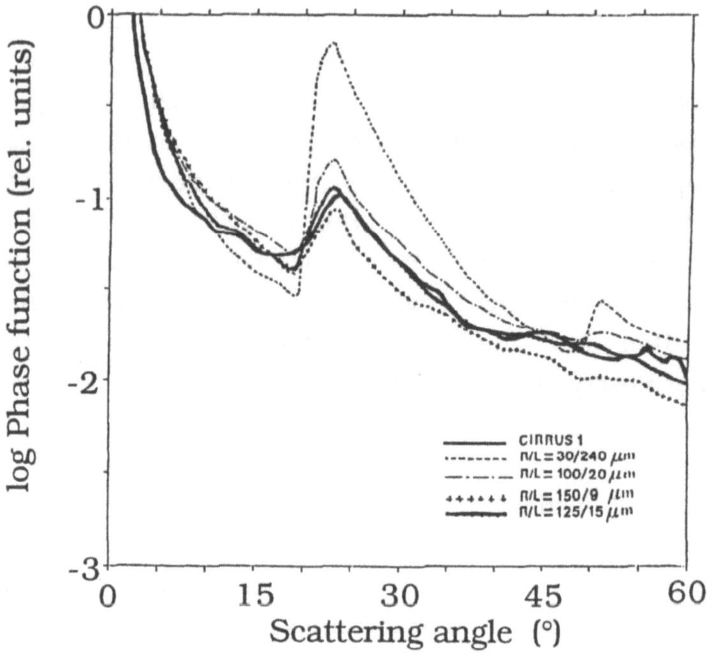

Fig. 3.3 Comparison of measured (CIRRUS 1) and calculated forward scattering during the ICE-89 experiment for the wavelength 0.85 μm. See text for details.

Hexagonal plates are generally characterised by two size parameters e.g., the thickness, L and the length of the hexagon, R. In the present case, the best agreement between measured and calculated angular distributions was reached for hexagonal plates with R = 125μm and L = 15μm. The bi-directional reflectance functions of cirrus clouds and the angular variation of the polarisation rate also differ significantly from those of liquid water clouds.

There has been evidence of preferential orientations of ice plates: (Platt, 1978) observed unusual LIDAR returns from altostratus clouds that he attributed to hexagonal ice plates falling with their long axis in the horizontal plane. Parhelions are also typical of oriented crystals. For the largest particles, this orientation is consistent with their aerodynamic properties.

Macke (1994) showed that crystal surface roughness could smooth out the halo phenomena very efficiently. The variability of crystal shapes and sizes as well as that of CWC is such that there is no unique model of ice crystals. There even is probably no unique model to represent particular cirrus clouds. Our problem here is thus to identify the specific optical properties which significantly affect cirrus observations and/or the role of cirrus in the radiation budget.

3.2 Bulk radiative properties of cirrus clouds

A first consequence of preferential orientations of crystals is that the extinction coefficient as well as the scattering phase function depend upon the direction of incident light. This dependence is the result of the variation of the cross section with the angle of incidence of radiation on the crystal faces. This cross section is obviously maximum for normal incidence of radiation. For identical volume, it is also larger for plates than for columns. The asymmetry factor is the first moment of the phase function. Its value is plus one for pure forward scattering and zero for isotropic scattering. (Masuda and Takashima, 1990) compared the extinction coefficients and asymmetry factors of spheres to those of columns and plates randomly oriented in space (3D) or in the horizontal plane (2D). For 2D plates with overhead sun, diffraction causes a large asymmetry factor with values close to 0.9. It is important to note that the asymmetry factor is significantly smaller for 3D columns than for spheres. This also holds for 2D columns at most zenith angles.

These radiative properties depend on wavelength through the variation of the relative contribution of diffraction. When absorption increases, the relative contribution of the diffraction peak increases and so does the asymmetry factor. At thermal infrared wavelengths, the geometric ray tracing techniques may be not accurate for small hexagonal crystals. This fact complicates the prediction of the asymmetry factor of ice crystals and spheres in this wavelength region. However, results of calculations performed by Kinne and Liou (1989) with this technique and Stephens (1980) using Discrete Space Theory indicate that the asymmetry factor of ice crystals is smaller than that of volume-equivalent spheres at these wavelengths. Consequently, in the thermal infrared, the reflectances of cirrus clouds are much larger than those of liquid water clouds. The absolute reflectance remains small but since the up welling radiation at cloud base is much larger than that emitted by the cloud itself, the radiation field above cirrus may be significantly affected.

3.3 The role of cirrus in the radiation budget

The effect of cirrus on the surface temperature is either a cooling or a warming depending on the predominance of their albedo- or greenhouse effect. The sign depends on the optical cirrus properties. (Stephens, et al., 1990) addressed this issue. With a very simple radiative equilibrium model they found that cirrus with asymmetry factor $g = 0.87$ warmed approximately twice as much as cirrus with $g = 0.7$.

Brogniez, et al (1994) evaluated the cloud radiative forcing of the particular cirrus that were observed during the ICE experiment. Using the model cloud that they had deduced from their forward scattering measurements they calculated the difference between short wave and long wave radiation fluxes for clear and overcast conditions at the top of the atmosphere, (cf. section 3.1.2). The results of these calculations are presented in Table 3.1 as a function of cirrus condensed water path for both their model cloud and for volume equivalent spheres.

Table 3.1 Short wave (CSW) and long wave (CLW) cloud forcing in Wm^{-2} for model clouds with spheres and 3-dimensional crystals having different condensed water paths (CWP) for the ICE-89 area in October 1989.

CWP	optical thickness		CSW		CLW	
gm^{-2}	spheres	crystals	spheres	crystals	spheres	crystals
1.2	0.026	0.1	-4	-5	5	18
6.1	0.13	0.5	-14	-18	23	66
24.4	0.73	2	-44	-56	83	113
121	2.63	10	-100	-122	116	118

For the particular conditions of the experiment, the influence of crystal shape is maximum for intermediate condensed water paths, reaching 40 Wm^{-2} for an optical thickness of 0.5. In a more general approach,

Stephens, et al. (1990) demonstrated that the sensitivity of their climate model depends on the crystal size and shape. They showed that even the sign of the cloud ice water-temperature feedback was uncertain.

3.4 Polarisation properties of cirrus

The polarisation signature of cirrus clouds has been studied by various authors: (Cai and Liou, 1982); Takano and Liou, 1989; Masuda and Takashima, 1992). The angular distribution of polarisation is compared in Fig. 3.4 to that for spheres. For the spheres an effective radius R_{eff} of 87.5μm was used resulting from a size distribution of the form

$$n(r) \approx r^6 \exp\left(\frac{6r}{r_0}\right) \quad (r_0 = 58\mu m) \quad .$$

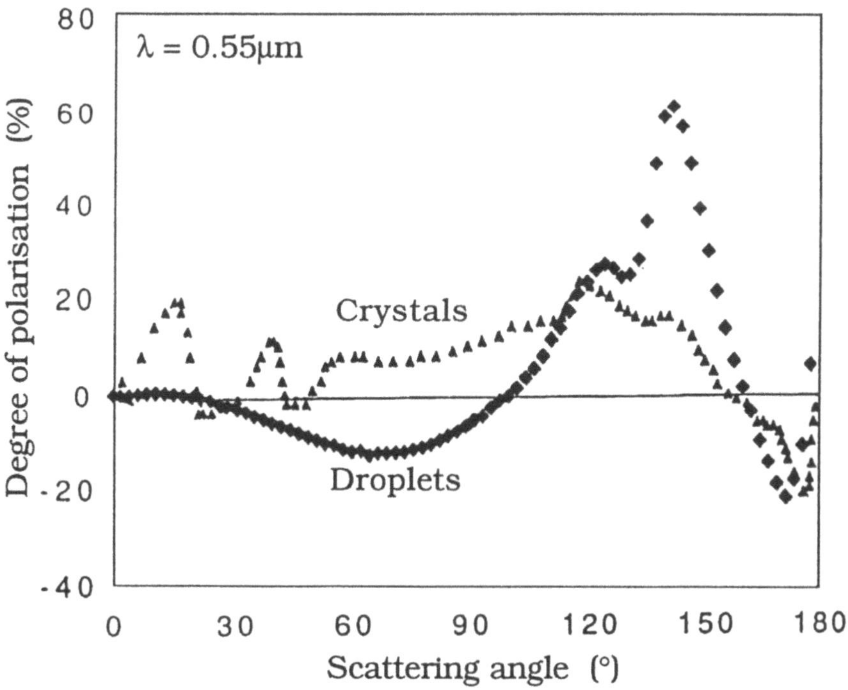

Fig. 3.4 Calculated degrees of polarisation for hexagonal ice crystals and spherical particles. See text for details.

The polarisation for these spheres is compared to that of a monodisperse size distribution of crystals having the same effective radius $R_{eff} = 87.5\mu m$ resulting from $R = 60\mu m$ and $L = 300\mu m$, (cf. section 3.1.2).

The degree of polarisation shows a distinct maximum near 150° corresponding to the rainbow. The exact location of the rainbow depends on the size of the scattering particles (Hansen and Travis, 1974). This polarisation maximum is absent for hexagonal ice crystals. For scattering angles below 160°, the degree of polarisation of hexagonal ice crystals is positive with exceptions for the halos. The location of the neutral points depends on the shape of the cloud particles.

These optical characteristics of crystals are expected to be detectable using specially designed polarimeters. Fig. 3.4 shows that a particular cloud (or scene in a satellite image) must be viewed from different angles to evaluate the angular dependence towards the shape of the hydrometeors. POLDER(Polarisation and Directionality of the Earth's Reflectances) is an instrument designed for that purpose. It is to be flown on the Japanese satellite ADEOS (ADvanced Earth Observation Satellite) in 1996. An airborne version was flown on the German research aircraft Falcon during the CLEOPATRA (Cloud Experiment Oberpfaffenhofen and Transport) experiment in 1991. No cirrus clouds were present during the observations. However, the first results show that the instrument is able to observe the polarisation of water drops.

3.5 Available methods for the treatment of radiative properties of non spherical particles

3.5.1 Single scattering parameters

Mie (1908) developed the theory of electromagnetic scattering by dielectric spheres of arbitrary size. Lord Rayleigh extended the theory to cylinders, Rayleigh (1918). Table 3.2 summarises the methods which presently are available for particular types of non spherical particles.

Table 3.2 Methods for the calculation of single scattering properties of non-spherical particles.

Method	Reference	Features
Electromagnetic theory: infinite cylinders	(Liou, 1972; Stephens, 1980b)	oriented and non oriented, exact solution
Electromagnetic theory: spheroidal particles	Asano and Sato (1980)	rigorous solution
Discrete dipole Approximation	(Purcell and Pennypaker, 1973; Flatau, et al., 1988)	small size parameters, all shapes
Anomalous diffraction theory	Van de Hulst (1957)	large size parameters
Geometric ray tracing	(Jacobowitz, 1971; (Wendling, et al., 1979; Van de Hulst, 1957; Coleman and Liou, 1982; Cai and Liou, 1982; Kinne and Liou, 1989)	hexagonal ice crystals, large size parameters

3.5.2 Multiple scattering in anisotropic media

Most methods for the solution of the equation of radiative transfer do not allow the calculation of multiple scattering in an anisotropic medium i.e., they cannot be applied to ice crystals with preferential orientations. Stephens (1980) used the Discrete Angles Approximation whereas Liou, et al. (1983), Masuda and Takashima (1990) and Minnis, et al. (1993) adapted the Doubling-Adding method to anisotropic scattering. Masuda and Takashima then explored the possibilities of deriving cloud characteristics such as phase and shape of particles from the degree of polarisation for clouds of varying optical thickness. They estimated the influence of the optical contributions of aerosols and of the earth's surface. They showed that the degree of polarisation should be measured with an accuracy better than ≈1% to infer the phase of hydrometeors. They also concluded that it would be necessary to have independent estimates of cloud optical thicknesses in order to infer the shape of the particles from the position of the neutral points in the angular distribution of polarisation. This position shifts to backward scattering angles with decreasing optical thickness because of the increasing influence of molecular and aerosol scattering and that of surface reflection. The most recently developed methods for calculating the radiative transfer through cirrus are listed in Table 3.3.

Table 3.3 Approximations and Parameterizations for the radiative transfer through cirrus clouds

Type of model	Reference	Usage
Two stream approximation; basic parameters expressed in terms of effective radius and asymmetry parameter	Stephens, et al., (1990)	broad band flux calculations
Schematic modelling of bulk cloud properties derived from doubling-adding for randomly oriented ice crystals; L/2a = 20/20, 85/40, 400/120 (μm)	Minnis, et al. (1993)	analysis of satellite observations
Four stream approximation; basic parameters extinction coefficient, single scattering albedo	Fu and Liou, (1993)	broad band flux calculations
Mie theory applied to surface-equivalent spheres scaled with the asymmetry factor g: g' = 0.85g; empirical correction for small particles based on Heymsfield and Platt (1984)	Sun and Shine, (1994)	broad band flux calculations

4. Open Issues and Questions

On time scales of an hour, geographically stationary cirrus structures have been identified by aircraft measurements well below 100% r.H. (over ice). These structures were recognised in a concurrent infrared satellite image. We do not understand yet this persistence of cirrus. The microphysical formation processes of cirrus are not yet understood either. How important is the liquid phase in these processes? Are bulk- and surface chemical processes on cirrus crystals affecting the evaporation of the ice crystals thus prolonging their life time? Are the so called sub visual cirrus clouds sub micrometer cirrus or sulphuric acid haze? There are no models available linking aerosol chemistry, cloud microphysics and cloud radiative transfer.

Presently and for some time to come, the calculation of the influence of cirrus on radiation in climate models is based on simple approximate methods that are more appropriate for the treatment of water clouds. Thus, the question of how to specify the basic parameters of these approximations is critical. Some attempts have been reported here but they are supported by a very limited set of data only. There is an strong need for more data for a large range of meteorological conditions. Satellite observations offer the possibility of global coverage. The issue is thus to relate cirrus signatures in satellite signals to quantities that can be predicted by models. Despite the attempts reported here, this issue essentially remains open.

References

Asano, S., and Sato, M. 1980. Light scattering by randomly oriented spheroidal particles. Appl. Opt. **19**, 962-974.

Brogniez, G., Buriez, J. C., Giraud, V., Parol, F., and Vanbauce, C. 1994. A radiatively equivalent microphysical model of cirrus cloud from ground-based and satellite observations during the International Cirrus Experiment. Mon. Weather Rev. (submitted).

Cai, Q., and Liou, K.-N. 1982. Polarized light scattering by hexagonal ice crystals: theory. Appl. Opt. **21**, 3569-3580.

Coleman, R. F., and Liou, K. N. 1982. Light scattering by hexagonal ice crystals. J. Atmos. Sci. **38**, 1260-1271.

Detwiler, A. 1989. Comments on "Homogeneous nucleation rate for highly supercooled cirrus cloud droplets. J. Atmos. Sci. **46**, 2344-2345.

Dowling, D. R., and Radke, L. F. 1990. A summary of the physical properties of cirrus clouds. J. Appl. Meteor. **29**, 970-978.

Flatau, P. J., Stephens, G. L., and Draine, B. T. 1988. *Scattering on hexagonal ice crystals: discrete dipole and anomalous diffraction approximations.* International Radiation Symposium, Lille, France, p.72-75.

Fu, Q., and Liou, K. N. 1993. Parameterization of the radiative properties of cirrus clouds. J. Atmos. Sci. **50**, 2008-2025.

Gardiner, B. A., and Hallett, J. 1985. Degradation of in-cloud forward scattering spectrometer probe measurements in the presence of ice particles. J. Atmos. Oceanic Technol. **2**, 171-180.

Heintzenberg, J. 1994. Properties of the log-normal particle size distribution. Aerosol Sci. Technol. (in print).

Heymsfield, A. J. 1986. Ice particles observed in a cirriform cloud at -83°C and implications for polar stratospheric clouds. J. Atmos. Sci. **43**, 851-855.

Heymsfield, A. J., and Platt, C. M. R. 1984. A parameterization of the particle size spectrum of ice clouds in terms of the ambient temperature and the ice water content. J. Atmos. Sci. **41**, 846-855.

Inoue, T. 1985. On the temperature and effective emissivity determination of semitransparent cirrus clouds by bispectral measurements in the 10 micrometer window region. J. Meteorol. Soc. Japan **63**, 88-98.

Jacobowitz, H. 1971. A method for computing the transfer of solar radiation through clouds of hexagonal ice crystals. J. Quant. Spectrosc. Radiative Transfer **11**, 691-695.

Kinne, S., and Liou, K. N. 1989. The effects of nonsphericity and size distribution of ice crystals on the radiative properties of cirrus clouds. Atmos. Res. **24**, 273-284.

Knollenberg, R. G. 1976. *Three new instruments for cloud physics measurements: the 2-D spectrometer, the Forward Scattering Spectrometer Probe, and the Active Aerosol Spectrometer.* International Conference on Cloud Physics, Boulder, CO, p.554-561.

Knollenberg, R. G., Kelly, K., and Wilson, J. C. 1993. Measurements of high number densities of ice crystals in the tops of tropical cumulonimbus. J. Geophys. Res. **98**, 8639-8664.

Krupp, C. 1991. *Holographic measurements of ice crystals in cirrus clouds during the International Cirrus Experiment Ice 89.* Fourth Workshop of the International Cirrus Experiment, Reading, U.K.,

Lin, X., and Coakley Jr, J. A. 1993. Retrieval of properties for semitransparent clouds from multispectral infrared imagery data. J. Geophys. Res. **98D**, 18501-18514.

Liou, K. N. 1972. Electromagnetic scattering by arbitrarily oriented ice cylinders. Appl. Opt. **11**, 667-674.

Liou, K. N., Cai, Q., Pollak, J. B., and Cuzzi, J. N. 1983. Light scattering by randomly oriented cubes and parallelepipeds. Appl. Opt. **22**, 3001-3008.

Macke, A. 1994. *Spectral variability of light scattering by atmospheric ice crystals*. In M. T. C. and N. A. S. A. Chedin (Ed.), *High spectral Resolution Infrared Remote Sensing for Earth's Weather and Climate Studies*. Springer, Berlin: p.191-204.

Masuda, K., and Takashima, T. 1990. Deriving cirrus information using the visible and near-IR channels of the future NOAA-AVHRR radiometer. Remote Sens. Environ. **31**, 65-81.

Mie, G. 1908. Beiträge zur Optik trüber Medien, speziell kolloidaler Metallösungen. Ann. Phys. **25**, 377-445.

Minnis, P., Liou, K.-N., and Takano, Y. 1993. Inference of cirrus cloud properties using satellite-observed visible and infrared radiances. Part I: Parameterization of radiance fields. J. Atmos. Sci. **50**, 1279-1304.

Noone, K. B., Noone, K. J., Heintzenberg, J., Ström, J., and Ogren, J. A. 1993. In-situ observations of cloud microphysical properties using the counterflow virtual impactor. J. Atmos. Oceanic Technol. **10**, 294-303.

Ogren, J. A., Heintzenberg, J., and Charlson, R. J. 1985. In-situ sampling of clouds with a droplet to aerosol converter. Geophys. Res. Lett. **12**, 121-124.

Parol, F., Buriez, J. C., Brogniez, G., and Fouquart, Y. 1991. Information content of AVHRR channels 4 and 5 with respect to the effective radius of cirrus cloud particles. J. Appl. Meteor. **30**, 973-984.

Platt, C. M. R. 1978. Lidar backscatter from horizontal ice crystal plates. J. Appl. Meteor. **30**, 973-984.

Platt, C. M. R., Scott, J. C., and Dilley, A. C. 1987. Remote sounding of high clouds. Part VI: Optical properties of midlatitude and tropical cirrus. J. Atmos. Sci. **44**, 729-747.

Purcell, E. M., and Pennypaker, C. R. 1973. Scattering and absorption of light by non-spherical dielectric grains. Astrophys. J. **186**, 705-714.

Lord Rayleigh 1918. The dispersal of light by a dielectric cylinder. Phil. Mag. **36**, 365-376.

Sassen, K., and Cho, B. S. 1992. Subvisual thin cirrus lidar dataset for satellite verification and climatological research. J. Appl. Meteor. **31**, 1275-1285.

Sassen, K., Griffin, M. K., and Dodd, G. C. 1989. Optical scattering and microphysical properties of subvisual cirrus clouds, and climatic implications. J. Appl. Meteor. **28**, 91-98.

Stephens, G. L. 1980. Radiative properties of cirrus clouds in the infrared region. J. Atmos. Sci. **37**, 435-446.

Stephens, G. L., Tsay, S.-C., Stackhouse, P. W., and Flatau, P. J. 1990. The relevance of the microphysical and radiative properties of cirrus clouds to climate and climatic feedback. J. Atmos. Sci. **47**, 1742-1753.

Ström, J., and Heintzenberg, J. 1993. Water vapour, condensed water and crystal concentration in orographically influenced cirrus clouds. J. Atmos. Sci. (in print).

Sun, Z., and Shine, K. 1994. Studies of the radiative properties of ice and mixed-phase clouds. Q. J. R. Meteorol. Soc. 111-137.

Takano, Y., and Liou, K.-N. 1989. Solar Radiative Transfer in Cirrus Clouds. Part I: Single- Scattering and Optical Properties of Hexagonal Ice Crystals. J. Atmos. Sci. **46**, 3-19.

Van de Hulst, H. C. 1957. *Light Scattering by Small Particles* . John Wiley and Sons, New York:

Wendling, P., Wendling, R., and Weickmann, H. K. 1979. Scattering of solar radiation by hexagonal ice crystals. Appl. Opt. **18**, 2663-2671.

Wu, M. C. 1987. A method for remote sensing the emissivity, fractional cloud cover, and cloud top temperature of high-level, thin clouds. J. Climate Appl. Meteor. **26**, 225-233.

Lord Rayleigh 1918. The dispersal of light by a dielectric cylinder. Phil. Mag. **36**, 365-376.

Sassen, K., and Cho, B. S. 1992. Subvisual thin cirrus lidar dataset for satellite verification and climatological research. J. Appl. Meteor. **31**, 1275-1285.

Sassen, K., Griffin, M. K., and Dodd, G. C. 1989. Optical scattering and microphysical properties of subvisual cirrus clouds, and climatic implications. J. Appl. Meteor. **28**, 91-98.

Stephens, G. L. 1980. Radiative properties of cirrus clouds in the infrared region. J. Atmos. Sci. **37**, 435-446.

Stephens, G. L., Tsay, S.-C., Stackhouse, P. W., and Flatau, P. J. 1990. The relevance of the microphysical and radiative properties of cirrus clouds to climate and climatic feedback. J. Atmos. Sci. **47**, 1742-1753.

Ström, J., and Heintzenberg, J. 1993. Water vapour, condensed water and crystal concentration in orographically influenced cirrus clouds. J. Atmos. Sci. (in print).

Sun, Z., and Shine, K. 1994. Studies of the radiative properties of ice and mixed-phase clouds. Q. J. R. Meteorol. Soc. 111-137.

Takano, Y., and Liou, K.-N. 1989. Solar Radiative Transfer in Cirrus Clouds. Part I: Single- Scattering and Optical Properties of Hexagonal Ice Crystals. J. Atmos. Sci. **46**, 3-19.

Van de Hulst, H. C. 1957. *Light Scattering by Small Particles* . John Wiley & Sons, New York, 470p.

Wendling, P., Wendling, R., and Weickmann, H. K. 1979. Scattering of solar radiation by hexagonal ice crystals. Appl. Opt. **18**, 2663-2671.

Wu, M. C. 1987. A method for remote sensing the emissivity, fractional cloud cover, and cloud top temperature of high-level, thin clouds. J. Climate Appl. Meteor. **26**, 225-233.

Lifetimes of Ice Crystals in the Upper Troposphere and Stratosphere

Th. Peter* and M. Baker**
*Max Planck Institute fur Chemie Postfach 3060 55020 Mainz, Germany

**Geophysics Program AK-50 University of Washington Seattle WA 98195 USA

1. INTRODUCTION

While it is increasingly apparent that small ice crystals in the upper troposphere and stratosphere may play important roles in radiative climate, quantifying these roles is very difficult. Until very recently the in situ measurements of crystal concentrations were made by optical instruments which are insensitive to particles of linear dimension lower than a few tens of microns (see, e.g., Heymsfield, 1973; Heymsfield and Platt, 1984; Sassen et al, 1989). These instrumental limitations have been overcome to a certain extent by introduction of new replicator techniques (Arnott et al, 1993) and the Counterflow Virtual Impactor (CVI) (Strom, 1993; Strom and Heintzenberg, 1994). The in-cloud findings of Arnott et al (1994) suggest that the radiative properties of cirrus may at times be dominated by those of the small particles. The new observations confirm earlier suspicions that there are many regions of the upper troposphere, outside clouds, in which there are relatively high concentrations of small ice crystals (linear dimension under 50 microns). Very recently, Strom and Heintzenberg (1994) found concentrations of around 10 - 100/L of crystals of linear size less than 30 microns which had been transported over wide regions of the upper troposphere over the Alps. In the stratosphere over Sweden, Deshler et al (1994) made the unexpected discovery of particles they determined to be ice, up to 5 microns in radius, at temperatures up to 10 degrees above the frost point. Thus the evaporation times for small atmospheric crystals appear to be long.

The first approximate calculations of evaporation rates date back to studies of ice crystals sedimenting from cirrus clouds or glaciated cumulonimbus anvils into an ice-undersaturated environment (Braham and Spyers-Duran, 1967; Hall

NATO ASI Series, Vol. I 35
Clouds, Chemistry and Climate
Edited by Paul J. Crutzen and Veerabhadran Ramanathan

and Pruppacher, 1977). The discovery that these particles may survive sedimentation over distances of several kilometers through highly undersaturated air led to the investigation of the seeder-feeder mechanism through which the precipitating particles may act as ice nuclei for low-lying clouds beneath, transforming them into ice clouds and possibly initiating storms with heavy precipitation (Braham, 1967). In calculations of the evaporation process Hall and Pruppacher assumed the particles to be of spherical or columnar shape with prescribed aspect ratio, and solved the ventilation-corrected gas and heat diffusion equations for the H_2O loss from the precipitating particles. Using this method the authors could not reach consistency with the observations of Braham and Spyers-Duran (1967), which might have been due to incomplete knowledge of temperatures and relative humitities below the cloud base; the observed lifetimes exceeded the calculated ones by more than a factor of three. In simulations of other observations by Heymsfield (1973), in which the meteorological information was more complete, Hall and Pruppacher (1977) had to assume a desorption coefficient $\alpha = 0.014$, corresponding to kinetic slowing of the evaporation due to a surface impedance, which basically reduced the purely gas-diffusion controlled evaporation rate by some 20 percent. However, observational evidence was and still is too scarce to decide whether the basis of these calculations is correct or other mechanisms must be considered.

In recent calculations of evaporation times of cirrus particles Strom and Heintzenberg (1994) also assumed the crystals to be spherical and to consist of pure ice with no surface impedance and found that the particles might persist for hours under the observed environmental conditions. Very different are the stratospheric measurements by Deshler et al (1994) which showed that ice crystals can survive more than hundred times longer than suggested by a simple model of evaporation of spherical particles. Peter et al (1994) invoked chemical effects on the surface of the particles to explain the longevity of these particles.

Because of the importance of realistic assessment of the particle lifetimes, we turn in the next section to some important physical and chemical processes that should be included in modelling vapor growth and evaporation (henceforth written g/e) rates of atmospheric crystals.

2. SHAPE AND SURFACE IMPEDANCE EFFECTS ON G/E RATES OF ICE PARTICLES

a. Background

Most existing models of ice crystal g/e in the atmosphere suffer from two important defects. The first is their inability to model realistic crystal shapes; in most calculations to date (Kuroda, 1984; MacKenzie and Haynes, 1993; Peter et al., 1994) the crystal shape is approximated as spherical or ellipsoidal whereas observations (Heymsfield and Platt, 1987; Sassen et al, 1989; Ramaswamy and Detwiler, 1986; Arnott et al, 1994) show that most ice crystals in cirrus are hollow or solid hexagonal columns or thick plates of linear dimension 20 - 500 microns and aspect ratio $\Gamma \equiv c/a = 0.1 - 10$, where c is the crystal length along the major and a the length along the minor axis. Similar shapes have been suggested for stratospheric ice particles by Goodman et al. (1989); however, this estimate applies only to the very largest particles collected at about 19 km altitude during the Airborne Antarctic Ozone Experiment. Detailed observations of stratospheric particles are scarce due to difficulties of access at those altitudes, the low concentrations and the small sizes of the particles.

The nonspherical shapes of crystals can have important effects on their radiative properties (Kinne and Liou, 1989) and g/e rates. However, the obstacles to computations presented by complex shapes and the lack of knowledge of the environmental determinants of crystal shape have traditionally led to calculations in which the complications of the shapes were ignored or, at best, replaced by quite arbitrary assumptions and rough approximations.

The second weakness in most computations of g/e rates for atmospheric ice crystals is their neglect of the complex surface processes that redistribute molecules landing on a crystal face. It is usual in these calculations to assume that the vapor density adjacent to the surface of a crystal is uniform; i.e., that

$$N_s = N_{eq}(T_s) \tag{1}$$

where $N(m^{-3})$ is the density of vapor molecules, the subscript 's' refers to the region just outside the crystal surface and the equilibrium value is that predicted by the Clausius-Clapeyron equation. The steady state vapor and heat diffusion equations are then solved in the environment outside the surface using Eq (1) as one of the boundary conditions. This assumption makes the computation of g/e rates analogous to the computation of the electrostatic potential surrounding a conductor with the shape of the crystal, and hence this model is called the 'capacitance' model (e.g., Pruppacher and Klett, 1978).

Eq (1) is not consistent with the observation that in steady state crystal faces advance or retreat pa allel to themselves in g/e, which requires that every point on a given crystal face receive or lose the same vapor flux. Assuming that diffusion on the surface does not redistribute molecules over dimensions comparable to the size of the crystal, this means

$$F = -D_g \underline{\nabla} N_s \cdot \underline{n} \tag{2}$$

is uniform, where F(*molecules* $m^{-2}s^{-1}$) is the vapor flux at the surface, $D_g(m^2 s^{-1})$is the diffusion constant of the vapor in air and n is the coordinate normal to the crystal face. (Note that for a sphere Eqs (1) and (2) are mutually consistent, whereas for nonspherical crystals they are not.) While it was recognized early that Eq (2) is the correct boundary condition for computations of crystal g/e rates, little attention had been paid to incorporating it into a g/e model until recently. The boundary condition on the vapor at large distances from the particle (assuming the particle is stationary) is

$$N \to N_\infty \text{ as } r \to \infty \tag{3}$$

In principle, a model built on Eqs (2) and (3) is straightforward. Laplace's equation for the vapor density in the medium outside the crystal must be solved subject to these boundary conditions. Uniformity of the flux renders the computation of the g/e rates analogous to the computation of the electrostatic potential outside a uniformly charged body. The boundary conditions for the temperature in the medium are entirely analogous to Eqs (2) and (3) for N. We assume, however, that the temperature on the surface changes very little over the surface, as is borne out by calculation.

Note that we are considering here ice particles growing directly from the vapor in the absence of liquid; in mixed phase clouds the sources for the heat and moisture diffusing to a growing crystal are supercooled drops, which complicates the diffusion problem considerably (Chen and Lamb, 1994).

Since the flux must go to zero as N_s approaches $N_{eq,s}$, the equilibrium value at the temperature of the particle surface, we also have

$$F = \alpha \frac{v_{th}}{4}(N_s - N_{eq,s}) \tag{4}$$

where $v_{th}(ms^{-1})$ is the thermal velocity of molecules in the vapor. Eq (4) defines the 'condensation (evaporation) coefficient' α, where $1/\alpha$ = surface impedance to absorption (desorption) of molecules from (to) the vapor. This impedance may be due to the presence of impurities on the surface as well as to scarcity of sites where an incoming molecule can bond or from which a crystal molecule can escape. Note that since F must be uniform over a crystal face, the local supersaturation

$$\sigma_s \equiv (N_s - N_{eq,s})/N_{eq,s} \tag{5}$$

varies over the face, implying that α is also a function of position on the face. (See Figure 1.) (We assume here that the temperature is uniform over the crystal surface.) This surface supersaturation can be found from Eqs (2) and (4) (and the heat diffusion equation); it is smaller in magnitude than, but proportional to, the environmental supersaturation σ_∞, where

$$\sigma_\infty \equiv \frac{N_\infty - N_{eq,\infty}}{N_{eq,\infty}} \tag{6}$$

Note that the supersaturation given by Eq (6) differs from that at the particle surface because both vapor density and temperature are different in the two locations.

Because the surface properties play an important role in the discussion to follow, we briefly summarize here two conceptual models for the crystal surface.

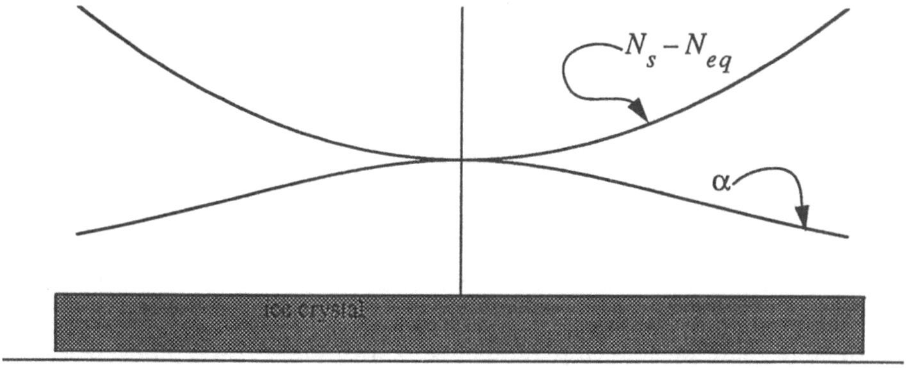

Figure 1: Schematic showing the variation of excess vapor density $(N_s - N_{eq,s})$ near a crystal face and the corresponding variation of the condensation coefficient (α) over that face. (From Nelson, 1994.)

b. The Crystal Surface

One of the earliest models of crystal growth is the so-called TLK (terrace-ledge-kink) model, in which during growth the crystal surface is considered to consist of flat terraces between straight, equidistant ledges, each one molecule high. Incoming vapor molecules land on the surface and diffuse over the terraces. As they diffuse some fraction (depending on the binding energy of the admolecules on the terrace) escapes back to the vapor. Some, on the other hand, remain on the surface long enough to eventually reach the nearest ledge. Reentrant corners, or kinks, in these ledges, are the most likely sites for absorption of incoming molecules, since these are the sites where new molecules can bind most strongly. Absorption of the molecules at the kinks advances the ledge and the growth rate of the crystal face perpendicular to itself is proportional to the ledge velocity. This conceptual model of the crystal surface was used by Burton, Cabrera and Frank (' BCF') (1951) to compute the rate of growth of a crystal face perpendicular to itself in terms of the surface supersaturation and the ledge spacing.

When the ledges are close together, the rate of incorporation of admolecules at kinks is high, whereas when the ledges are far apart the incorporation rate is much slower. That is, α is low at large ledge spacing and rises as the ledge spacing falls. The ledge spacing, in turn, is determined by the ledge creation mechanism, which is usually thought to be either nucleation (Frank, 1982) or spiral dislocations (BCF). In these cases, the ledge spacing is a decreasing function of surface supersaturation. Thus this model gives low values of α at low σ_s and α rises steeply to 1 as σ_s increases. (See Figure 2.) This model can be applied to evaporation if the ledge generation mechanism is known.

A second model for the crystal surface is suggested by recent experimental and theoretical studies on surface melting. (See, for example, Dash (1989) for a review of this subject.) Surface melting is the formation of a thin layer of melt liquid, or a so-called 'quasiliquid' layer (QLL) on the freshly cleaved surface of a solid at a temperature below its bulk melting point. The condition that surface melting occurs at a solid/vapor interface can be understood by imagining a block of solid material at a temperature below the bulk melting point, where the solid is the only bulk phase in equilibrium with vapor. Creation of a fresh surface of area A (by cleaving the block, for example) changes the free energy of the system by an amount $\bar{\gamma}_{sv} * A$, where $\bar{\gamma}_{sv}(Jm^{-2})$ is the nominal surface energy of the solid against vapor. If

$$\bar{\gamma}_{sv} \geq \gamma_{sl} + \gamma_{lv} \tag{7}$$

(where γ_{lv}, γ_{sl} are the surface energies of the liquid against vapor and solid against liquid, respectively) then the free energy of the system can be minimized by forming a thin melted layer on the surface. Note that Eq (7) is the classical condition that the melt 'wet' the solid.

The change in the free energy of the system as the bare solid/vapor interface is replaced by a solid/surface melted/vapor interface is a function of the surface energies and of μ_l and μ_s, the the chemical potentials in the liquid and solid phases. (Here we assume the thermodynamic properties of the surface melted material are identical to those in the bulk melted state.) The two interfaces of the thin QLL interact, so that the surface part of the free energy is a function of the layer thickness. Minimizing the free energy of the system gives the equilibrium thickness of the layer, $h(T,p)$, at which the interaction of the two interfaces stabilizes the film outside the temperature regime in which the bulk liquid and solid can coexist. In general, $h(T,p)$ is no more than a few molecular diameters at temperatures more than a few degrees below the bulk melting temperature.

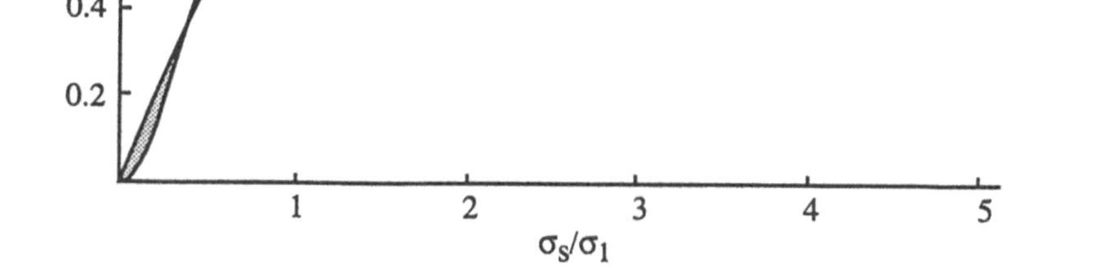

Figure 2: Condensation coefficient vs surface supersaturation: spiral ledges. The two curves represent the diffusion regimes in the lower and upper atmosphere. The parameter σ_1 must be fit from laboratory experiment. (From Nelson, 1994).

Surface melting at an ice/solid interface is well established experimentally and theoretically. At the ice/ pure water vapor interface, partial surface melting is expected on theoretical grounds for temperatures near the triple point (Elbaum and Schick, 1992) but the experimental evidence on surface melting at the ice/air interface at lower temperatures is widely varying.

If ice crystals in air surface melt, $1/\alpha$ would represent the impedance of the surface melted layer to absorption of molecules from the vapor. The relationship between α and local supersaturation in this case has not yet been explored theoretically and there are very few laboratory results which are pertinent, so we do not discuss this picture further.

c. G/e rates

We have solved the diffusion equations for heat and vapor subject to boundary conditions Eqs (2) and (3) for crystals in the form of finite cylinders (as well as spheres and ellipsoids) under the assumption that there is one point where ledges are generated on each crystal face, and that the BCF model for the relationship between α and supersaturation applies at that point. This allows us to find the flux at that point on each face, and therefore everywhere on the face. (At other points on the face the surface supersaturation is determined as the solution to the steady state diffusion equations. This causes the ledge spacing to vary over the crystal face and thus α also varies, but, as required by Eq (2), the product $\alpha\sigma_s$ is constant. See Figure 1.) Under the assumption (supported by observation (Frank, 1982)) that the ledges are often generated at the crystal corners this yields the following expression for the steady state g/e rate of growth of the a face of a cylindrical crystal of aspect ratio Γ in an environment at temperature T and environmental supersaturation (undersaturation) σ_∞.

$$\frac{da}{dt} = \frac{m_v\,(v_{th}/4)\,N_{eq,\infty}\sigma_\infty}{\rho(1/\alpha_a + a'(h_a + \Gamma^2 h_c))} \tag{8}$$

where m_v is the molecular mass in the vapor, ρ the density of the crystal, h_a and h_c are tabulated functions of c and a, and $a' \equiv a v_{th}/(4D_g)$. $1/\alpha_a$ is the impedance on the surface of the 'a' face. (A symmetric expression could be written in terms of the parameters of the 'c' face.) We have assumed here that in steady state growth $\alpha(a)/\alpha(c) = \Gamma$, and we have neglected the heat flux into the particle. This model has been termed a pseudocoupled model and its full derivation is given in Nelson (1994) and Nelson and Baker (1994).

The form of da/dt in Eq (8) shows that the g/e rates are driven by the supersaturation (undersaturation) in the crystal environment, and that the rates are slowed by (i) the surface impedance, parameterized by the term $1/\alpha$, and (ii) the thermal and diffusion impedances to the flow of vapor in the medium. The relative importance of these impedances is governed by the product $\alpha a'$. In the case

$$\alpha a' << 1$$

the surface impedance is more important, while for

$$\alpha a' >> 1$$

the volume impedances limit the g/e rates. Clearly, if the particles are small, and/or the super/undersaturation in the environment is small, the surface processes are the limiting factors in g/e. On the other hand, if the super/undersaturation is very high (as in mixed phase clouds) according to the TLK model of the crystal surface the ledges are close together (due to high σ_s) so that $\alpha \rightarrow 1$, and the surface impedance becomes negligible compared to the volume impedances. The capacitance model is equivalent to eq (8) in the limit $1/\alpha \rightarrow 0$, although in general the capacitance model has not been applied to finite cylinders, but rather only to needles, infinitely thin discs and ellipsoids. The shape effect can be extremely important; cylindrical columns grow more slowly than ellipsoids of the same aspect ratios, while for Γ approximately 1 ellipsoids grow more slowly than finite cylinders. The difference in growth rates is about a factor of ten for zero surface impedance, $\Gamma \approx 10$.

Under the conditions prevalent in fully glaciated clouds, or the slightly undersaturated environments around clouds, the surface impedance may become the factor limiting growth. Under these conditions the pseudocoupled model is in principle more accurate than the capacitance model. In order to apply it we must specify α as a function of surface super/undersaturation. Little is known experimentally about the ice surface in this regard. If we assume that the surface impedances are not due to impurities, but are due to diffusion processes on the ice surface, we can use models for $\alpha(\sigma_s)$ such as the BCF model described above, and we can fit these to laboratory studies to infer the values of the numerical parameters that enter those models. In general the predicted g/e rates for the pseudocoupled model are lower than those predicted by the capacitance model. Figure 3 shows, for instance, the ratios of the two mass uptake rates for a range of values of the ratios of surface to diffusive impedance. For large values of this ratio the capacitance model overestimates the mass uptake ratio considerably, while for relatively small surface impedance the two models give approximately the same results. Figure 4 shows the ratios of the linear growth rates predicted on the basis of the capacitance models and the pseudocoupled models for a range of parameters (used to evaluate the surface impedance) inferred from laboratory data, and a range of atmospheric pressures.

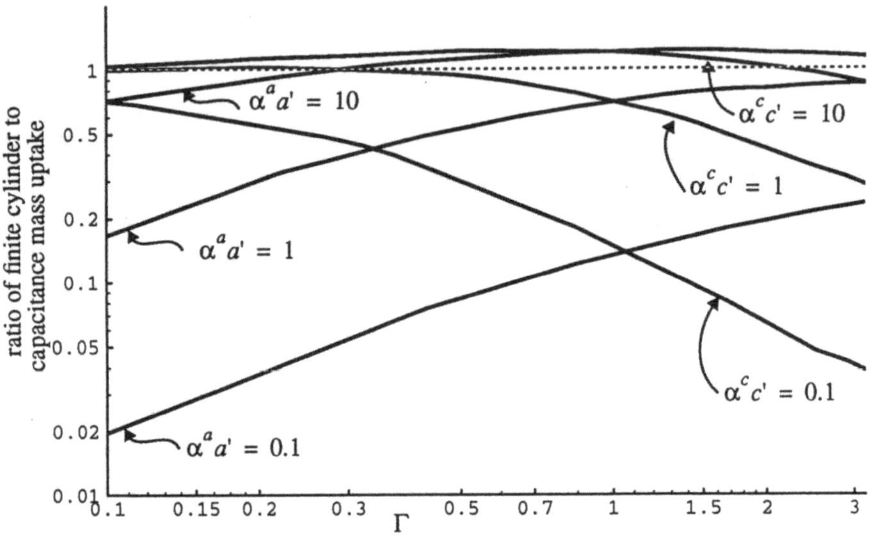

Figure 3: Ratio of the mass uptake rate of a finite cylinder (from pseudocoupled model) to that predicted by the capacitance model for an ellipsoid of the same aspect ratio and no surface impedance. α^a, α_c are the condensation coefficients on the sides and base of a finite cylindrical crystal, while a', c' are the nondimensionalized linear dimensions of those faces. See text. (From Nelson and Baker, 1994.)

The capacitance model is symmetric for growth and evaporation. Therefore the g/e rates predicted by it are equal at equal departures from equilibrium. If, however, the surface processes involved in evaporation and growth are not the same, the surface impedance in Eq (8) will not be symmetric about $\sigma_s = 0$ and in principle the magnitudes of the two rates can be very different at equivalent under- and supersaturations. This might be the case, for example, if evaporation removes molecules preferentially from certain sites (such as corners), whereas molecules are incorporated elsewhere preferentially during growth.

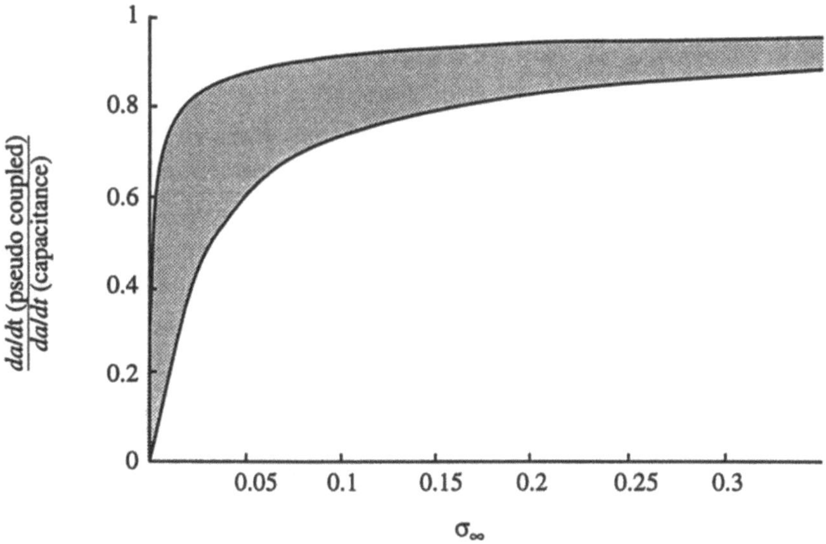

Figure 4: Ratio of the linear growth rates predicted by the pseudocoupled and capacitance models for $\Gamma = 1$, and a range of surface parameters derived from laboratory experiments. Based on Nelson and Baker, 1994.

In the discussion so far we have not mentioned the effects of impurities on g/e rates. We consider here impurities in the gas phase (i.e., environmental constituents other than water vapor) which may be incorporated into the lattice and/or simply adsorbed on the surface of a crystal. Of particular interest in the upper troposphere are H_2SO_4 and HNO_3. Analysis of the particle residues obtained by evaporating the cirrus crystals detected by the CVI in flights over the Alps (Strom et al, 1994) showed that the mass mean radius of the residues was around 0.2 microns. If we assume the particles are sulfates, then the crystals which nucleated on them could be covered by patches of sulfuric acid solution which could possibly alter their g/e rates. In the stratosphere, as discussed below, gas phase HNO_3 may cocondense with water or it may form a NAT coating on pure ice crystals which both might considerably change g/e rates.

The effects on g/e rates of impurities incorporated into the crystal lattice as well as those simply adsorbed on the surface are to modify several terms on the RHS of eq (8). First, the existence of a foreign gas in the crystal environment modifies the diffusion of water vapor and heat through the medium, and therefore it can modify the concentration of water vapor molecules at the surface.

These are the main effects of nitrogen and oxygen and the other "air" molecules. Trace gases, whether or not they are soluble in water, can also alter the crystal surface properties, as represented by α. These effects seem to be strongest for molecules with high dipole moments, which may be those molecules that bind most effectively to ice (Nakaya et al, 1958). There is, at present, no theory relating molecular structure of a compound to its effect on α. Impurities may also greatly alter the propensity of ice to surface melt by changing the chemical potentials of the vapor and liquid (and, to some extent, the various surface energies). Estimates based on dilution solution theory of the thickness of the surface melted layer as a function of temperature and chemical composition (B. Swanson, personal communication) yield some enhancement of surface melting at lower temperatures. However, laboratory results on the uptake of impurities by ice at temperatures well below the triple point (e.g., Clapsaddle and Lamb, (1989), Valdez and Dawson (1989), Conklin et al, (1993), Molina (1992)) seem to indicate the presence of far more liquid on the ice surface than can be explained in this way. There is at present no theoretical explanation of the laboratory results.

3. MODELLING EFFECTS OF CHEMICAL COATING ON G/E RATES OF ICE CRYSTALS

In Section 2 we investigated the effects of crystal shape and surface impedance on g/e rates of ice crystals. In the discussion above impurities were considered only as trace substances which may alter the diffusion of water molecules on the crystalline terraces and the absorption of water molecules at crystal ledges and kinks, the active sites to which the water molecules absorb or from which they desorb. There is no laboratory evidence which shows that due to such mechanisms atmospheric concentrations of impurities have a significant influence on g/e rates of tropospheric ice crystals (Nelson, 1994). We did not discuss possible incorporation of impurities into the body of a crystal. In the stratosphere it is possible that ice crystals actually absorb some impurities at high enough concentrations that these form coatings whose equilibrium vapor pressure is far lower than that of ice.

Under average stratospheric conditions (T = 210--230 K) water partial pressures are much too low for the formation of ice particles. For example, at 50 hPa (ca 20 km altitude) with 5 ppmv of H_2O (typical at those altitudes), the frost point is 188 K, a temperature which is reached only under the coldest polar winter conditions. However, there is another crystalline compound which has a lower vapor pressure than ice under stratospheric conditions: nitric acid trihydrate ($NAT \equiv HNO_3x3H_2O$) . Assuming typical values of 10 ppbv HNO_3 and 5 ppmv H_2O at 50 hPa, NAT has a thermodynamic equilibrium temperature of 195 K; i.e. it can exist at temperatures 7 K above the frost point. Clearly, the stratospheric air encounters this higher temperature much more often than the frost point. Hence clouds containing NAT or other nitric acid compounds are sighted more frequently than stratospheric ice clouds (WMO, 1991). On the other hand, since the mixing ratio of H_2O in stratospheric air is about 500 times greater than that of HNO_3, nitric acid containing clouds develop only small particles (typical linear dimensions$< 0.5 \mu m$) at number densities up to $10\ cm^{-3}$) while stratospheric ice crystals can become quite large (ranging from a few μm up to more than 100 μm) at concentrations of $0.001$to $10\ cm^{-3}$ depending on supersaturations, nucleus densities and growth times (cp. Goodman et al., 1989).

The interest in the thin veils of polar stratospheric clouds (PSCs) has bloomed since their importance for ozone chemistry became evident (Solomon et al 1986). For both NAT and ice particles there is ample laboratory as well as observational evidence that important chemical reactions of nitrogen and chlorine species take place on the particle surfaces; indeed, without these heterogeneous reactions the annual development of the Antarctic ozone hole can not be explained. While the reaction probability (WMO, 1991) for the dominant path of chlorine activation

$$ClONO_2 + HCl --- > Cl_2 + HNO_3$$

is similar on both NAT and ice surfaces, an important difference between these species is that the larger ice particles sediment out of the stratosphere faster than NAT particles, thereby eliminating water and other absorbed species from the stratosphere.

In the Antarctic stratosphere the observed denitrification (loss of NO_x and HNO_3) leads to a considerable perturbation of gas phase chemistry in favor of enhanced ozone loss. Although this has been attributed to the sedimentation of ice particles growing on NAT particles, the observed denitrification can not easily be explained by this mechanism alone because ice particle number densities are much lower than those of NAT particles. Wofsy et al. (1990) hypothesized that ice particles could sediment into warmer regions --- above the frost point for ice but below that for NAT --- and accrete a coating of NAT that effectively inhibits evaporation of the ice particles. This hypothesis, combined with the fact that NAT particles tend to supercool by 3--4 K before they nucleate (Schlager et al., 1990), leads to the suggestion that ice particles may sediment into regions where the full HNO_3 reservoir is still available to provide the coatings. For the example p = 50 hPa, 5 ppmv H_2O, 10 ppbv HNO_3, and T = 191 K we find that one 3.5-micron ice particle cm^{-3} could acquire a NAT coating of about 20 nm. NAT coatings of similar thicknesses have been shown to effectively prevent ice evaporation in laboratory experiments (Tolbert and Middlebrook, 1990).

Some of the few stratospheric observations of large ice particles at temperatures apparently too warm to sustain them were made above Finland on 27 January 1992 using a balloon-borne aerosol counter (Deshler et al., 1994). On this flight temperatures were even slightly above the NAT equilibrium temperature, except around 20 km altitude. The particles were found to be very finely layered in the vertical which indicates their probable origin in patches of cold air induced by leewaves above the Norwegian mountains. In the highly undersaturated environment $(T \approx 195K, \sigma_\infty \approx -0.66)$ downstream of the mountains, calculations neglecting surface effects (i.e., using Eq (8) with $\alpha a' >> 1$) show that pure ice particles could have survived for only a few minutes, while NAT particles would endure for hours. However, if the particles contain both ice and nitric acid, condensed simultaneously during rapid cooling, a coating of NAT may form during evaporation from the HNO_3 molecules which are probably embedded in the ice crystal in the form of NAT clusters during condensation. As suggested by Peter et al. (1994) these clusters could accumulate at the particle surface during the evaporation of the ice and may form a closed NAT-coating, preventing further evaporation if $T < T_{NAT}$, or, if $T > T_{NAT}$, considerably reducing the vapour pressure of the evaporating particle.

For stratospheric ice crystals it is not clear *a priori* which processes limit evaporation rates. On the one hand, particle radii a are small and the diffusion constant D_g is large due to the low pressure. Hence a' is small. On the other hand, supersaturations σ_∞ in the stratosphere can become much larger for longer periods of time than in the troposphere. This is because the decay time for the supersaturation is determined by the particle g/e rates, which are proportional to the particle concentrations and the equilibrium vapor density $N_{eq,\infty}$ (see Eq 8). Stratospheric values for the particle concentrations and for $N_{eq,\infty}$ are typically 2--4 orders of magnitude lower than tropospheric values. Temperature changes in air parcels are therefore likely to be associated with much larger supersaturations in the stratosphere than in the troposphere. For orographically driven temperature waves brief excursions of σ_∞ of more than 20% are predicted by model calculations. This in turn leads to large surface supersaturations, and thus large $\alpha(\sigma_s)$. On the other hand, for very small crystals there may be only a few ledges on the surface. Under these conditions the surface impedance becomes much more important than implied by Figure 2, which was based on the assumption that there were many ledges. (Nelson, 1994). It is therefore not clear whether the shapes, the ice surface, possible NAT coatings, or the volume impedances limit the g/e rates in the stratospheric case. We assume for the present calculations that the volume impedances and the presence of NAT are limiting and that the particles are spherical. In this case for $\alpha = 1$ for evaporation of a single gas Eq (8) becomes

$$\frac{da}{dt} = \frac{v_{th}/4m_v N_{eq,\infty}\sigma_\infty}{\rho(1+a')} = \frac{p_{eq}D'_g m_v \sigma_\infty}{\rho kTa} \equiv \frac{\beta\,\sigma_\infty}{a}, \tag{9}$$

where p_{eq} is the equilibrium vapor pressure and ρ the density of the condensed phase (NAT or ice). $p_{eq} = N_{eq,\infty}kT$, where k is the Boltzmann constant, and D'_g is the modified gas phase diffusion constant of the condensible (HNO_3 *or* H_2O); $D'_g \equiv D_g/(1+1/a')$. Eq (9) ignores effects of heat diffusion, radiation and ventilation, which for small particles at cold stratospheric temperatures is reasonable. In the region of interest (T = 190 K, p = 50 hPa) the growth coefficient β is approximately

$$\beta = \frac{1.1x10^4(p_{eq}/p)(T/190K)^{0.94}}{(1+1.6/a)}\,(microns^2 s^{-1}) \tag{10}$$

for both NAT and ice. Assuming $\beta\sigma_\infty$ remains constant during the evaporation solution of Eq (9) yields the following characteristic times for halving or doubling the initial particle radius a_0 (Toon et al., 1989):

$$\tau_{1/2} = \frac{3a_0^2}{8\beta\,|\,\sigma_\infty|}, \; \tau_2 = 4\tau_{1/2}. \tag{11}$$

In warm regions with $\sigma_\infty \to -1$, the lifetime of NAT particles is much longer than for an ice particle at the same temperature since $p_{eq}(NAT) \ll p_{eq}(ice)$. As Table 1 shows, the lifetime for an ice particle with $a_0 = 1\mu m$ at T = 192 K, p = 50 hPa is less than a minute, while that of a NAT particle is almost 10 hours (assuming 5 ppmv H2O).

The question what happens when HNO_3 molecules co-condense during ice growth under stratospheric conditions is still open to scientific debate. The HNO_3 molecules might be built as impurities into the ice crystal where they form single NAT units. From Eq.(8) (applied to both vapors) we estimate the NAT mole fraction in ice can reach $f \equiv HNO_3/H_2O = 1/500$. As outlined above, when temperatures rise this mixture of ice and NAT could evaporate until a monolayer of NAT has formed. Then, if the temperature is below T_{NAT} the evaporation of NAT will stop, and subsequent NAT condensation will seal the ice inside.

The proposed coating mechanism is similar to the one envisaged by Wofsy et al. (1990), but instead of sedimentation of the ice particles into warmer regions in the case described we imagine the entire air parcel warming up after traversing the Scandinavian mountains. If the air warms up quickly to temperatures above T_{NAT}, then after evaporation of the first NAT monolayer, ice is again exposed to the atmosphere and the cycle repeats. NAT would thus slow, but not stop evaporation. In an idealized picture Eq (8) has then to be solved successively for the evaporation of many layers of ice (roughly 500/3) followed by one layer of HNO_3, and so on. We can find the evaporation rate for the mixture as follows. We assume the NAT is distributed uniformly throughout the crystal and successively closed surface monolayers of ice and NAT are formed during the process of evaporation. Moreover, we assume that these layers develop the same vapor pressure as bulk ice or NAT. Then the time to evaporate N_{NAT} layers of NAT (each of thickness Δ_{NAT}) and N_{ice} layers of ice (each of thickness Δ_{ice} is

$$\Delta t = [\frac{N_{NAT}\Delta_{NAT}a}{\beta_{NAT}\sigma_{\infty,NAT}} + \frac{N_{ice}\Delta_{ice}a}{\beta_{ice}\sigma_{\infty}}] \tag{12}$$

In this time the radius grows by an amount $\Delta a = N_{NAT}\Delta_{NAT} + N_{ice}\Delta_{ice}$, and hence in the limit Δt, $\Delta a \to 0$ Eq (12) takes the form

$$dt = \frac{ada}{\beta_{mix}\sigma_{\infty}} \tag{13}$$

where

$$\frac{1}{\beta_{mix}} = \frac{f}{\beta_{NAT}}\frac{\sigma_{\infty}}{\sigma_{\infty,NAT}} + \frac{(1-f)}{\beta_{ice}} \tag{14}$$

with

$$f \equiv \frac{N_{NAT}\Delta_{NAT}}{(N_{ice}\Delta_{ice} + N_{NAT}\Delta_{NAT})}$$

For highly undersaturated regions (i.e., when most of the gas phase HNO_3 is bound up in the particles $\sigma_\infty \approx \sigma_{\infty,H_2O} \approx -1$. Thus

$$\frac{1}{\beta_{mix}} \approx \frac{f}{\beta_{NAT}} + \frac{(1-f)}{\beta_{ice}} \tag{15}$$

In these expressions $f = 0$ for pure ice and $f = 1$ for pure NAT. For the leewave-generated particle observed by Deshler et al (1994) we estimate $f \approx 0.003$.

According to Eq (11), for high supersaturations the time for the radius of a spherical particle to diminish from $1\mu m$ to $0.5\mu m$ via evaporation is approximately

$$t' \approx 3\mu m^2/(8\beta_{mix}\,|\,\sigma_\infty\,|).$$

In Table I we show the quantities t' for particles of pure ice, pure NAT and a mixture of the two. The quantities in the last column of Table I were computed according to the model outlined above, and reveal that such mixed crystal structures might survive 10 - 100 times longer than pure ice particles in a strongly undersaturated environment. This effect, of course, strongly depends on the assumptions made above, which essentially mean that the vapor pressure changes linearly with the solid state mole fractions. This is probably an overestimation, since it has been assumed that individual NAT clusters have the same vapour pressure as bulk NAT. Experimental investigation of this question would be helpful.

TABLE 1. Temperature dependence of the lifetime of particles with $1\mu m$ radius in a highly subsaturated environment at p=50 hPa. See Eq (11). For NAT we assumed 5 ppmv H_2O. The times in column 4 refer to water ice with a mixing ratio 1/500 HNO_3 as impurity. Lifetimes scale roughly with the square of the radius (Eq. 11). Vapour pressures are from Hanson and Mauersberger (1988).

$T(K)$	t'_{ice}	t'_{NAT}	t'_{mix}
185	72s	2.1a	55h
190	30s	16d	68m
192	22s	3.5d	15m
195	13s	9.5h	116s
200	6.1s	17m	9.1s
205	2.8s	34s	2.9s
210	1.3s	1.4s	1.3s

The suggested protection of the ice by a NAT layer also assumes that the solid phase diffusion of H_2O through NAT (and/or diffusion of NAT through H_2O) is negligible. Solid phase diffusion constants are around $10^{-16}m^2s^{-1}$. If this diffusion occurred even a 20 nm layer of NAT should not be able to effectively protect the ice since the diffusion time for water through the layer$I\ pprox$ $(2x10^{-8})^2/10^{-16} = 4s$, in contrast to experimental evidence (Middlebrook and Tolbert, 1990). One possible explanation for this could be that the NAT coating conserves the $H_2O - H_2O$ bonds in the ice and suppresses solid phase diffusion from the beginning. This is one of many points which should be investigated in the future.

Figure 5 shows the two mechanisms suggested for coating: the slow mechanism of condensation-coating due to ice particle sedimentation into warmer air (Wofsy et al., 1990) and the fast formation process due to NAT-flaws growing together during leewave-induced evaporation. Another difference between these two mechanisms is the fate of the sulfuric acid core from which the ice particles originate. As has been shown by Dye et al. (1992) H_2SO_4 aerosol particles may remain liquid to very low temperatures, inhibiting straightforward NAT nucleation. However, below the frost-point ice crystals may grow within the solution, while the growth of the crystals is essentially undisturbed by the surrounding liquid $H_2SO_4/HNO_3/H_2O$ layer (Luo et al., 1994). Hence, it was suggested that the mechanism of NAT-doped ice formation starts from a liquid aerosol droplet instead of a solid NAT core (see Fig.5). In this process there is no formation of NAT particles prior to ice nucleation, and so the gas phase HNO_3 reservoir is not reduced but remains available for uptake into the growing ice.

For a simulation of the ice particle observations on 27 January 1992 a comprehensive microphysical PSC-model was used which treats the following processes: nucleation of NAT on frozen aerosol and of ice on NAT based on classical nucleation theory; condensation and evaporation of NAT and ice; co-condensation of HNO_3 into growing ice; and, in addition, NAT-coating of ice particles and ice nucleation in liquid aerosol particles. This model was forced by time dependent temperature and pressure fields derived from mesoscale trajectories.

Volkert and Intes (1993) showed that synoptic (large-scale) trajectories often seriously misrepresent stratospheric temperatures in orographically perturbed regions. On 27 January 1992 strong tropospheric winds induced pronounced lee-

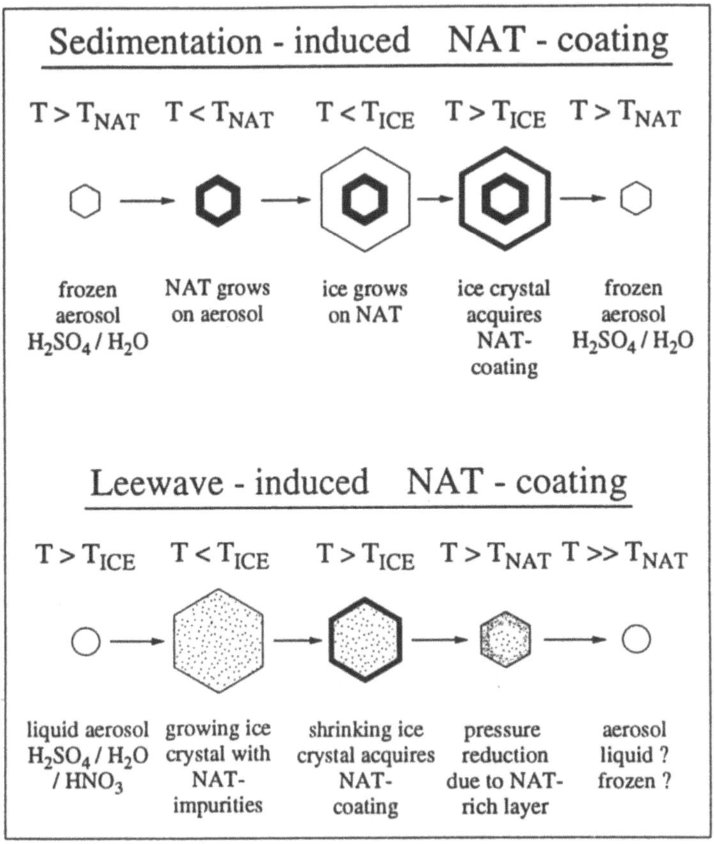

Figure 5: Upper panel: Sketch of the NAT coating mechanism proposed by Wofsy et al. (1990). Ice particles sediment into warm regions where NAT particles have not yet formed and acquire a NAT coating which protects them against evaporation. Lower panel: Sketch of the evolution of an ice crystal with NAT-impurities nucleating in a liquid H_2SO_4/H_2O aerosol particle. In this case ice evaporation is slowed due to a NAT-coating ($T_{ice} < T < T_{NAT}$) or due to a lowering of its vapour pressure ($T > T_{NAT}$).

waves over the Norwegian mountains. Without this mesoscale effect temperatures wouldn't even drop below T_{NAT}. The temperature and pressure history of an air parcel passing through the mountaineous region is shown in Figure 6. The detailed structure (Volkert and Intes, 1993) is corrected to force agreement with upstream (radiosonde data from Bodö) and downstream (Deshler et al., 1994) observations. The accuracy of the leewave calculation is limited by its 2-D character and the smoothing of the orography (lower panel, Figure 6). Figure 7 shows

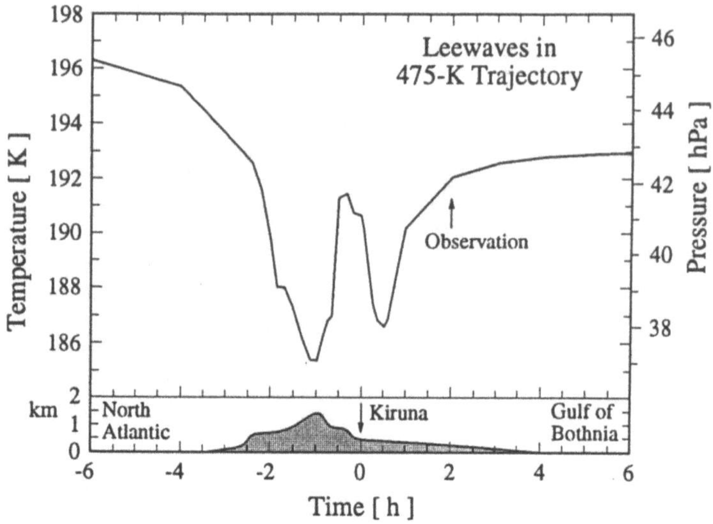

Figure 6: 12-hour trajectory (Volkert and Intes, 1992) at 475 K over Scandinavia. Time zero is at Kiruna. The smoothed orography for the 2-D leewave model is shown below.

the results of the model runs. Three different cases are presented: (1) no coating/solid aerosol (left); (2) NAT-coating/solid aerosol (center); (3) NAT-coating/liquid aerosol (right). For each case calculated particle concentrations (top row) and mean radii (center row) are plotted as functions of time (corresponding to the time axis in Figure 6) and the radial distributions are shown in the bottom row. All three runs use 5.5 ppmv H_2O and 9 ppbv HNO_3 and an initial aerosol density n=12 cm^{-3} with log-normal size distribution (mode radius $a_m = 0.07 \mu m$, width $\sigma_a = 1.9$).

*Case*1: No coating. In this case there is a short flowering of ice particles which lasts for about an hour. The mean radius of the remaining NAT particles ($\bar{a}_{NAT} = 0.3 \mu m$) is much too small to explain the observed large particles (diamonds, lower panel Figure 7).

*Case*2: NAT-coating, frozen background aerosol. NAT formation starts at t=-2 h (two hours before reaching Kiruna) and consumes about 7 of the 9 ppbv HNO_3, until at t=-1.2 h ice formation exhausts the other 2 ppbv due to co-condensation. Before t=+0.5 h there is NAT supersaturation and a coating can be achieved (the

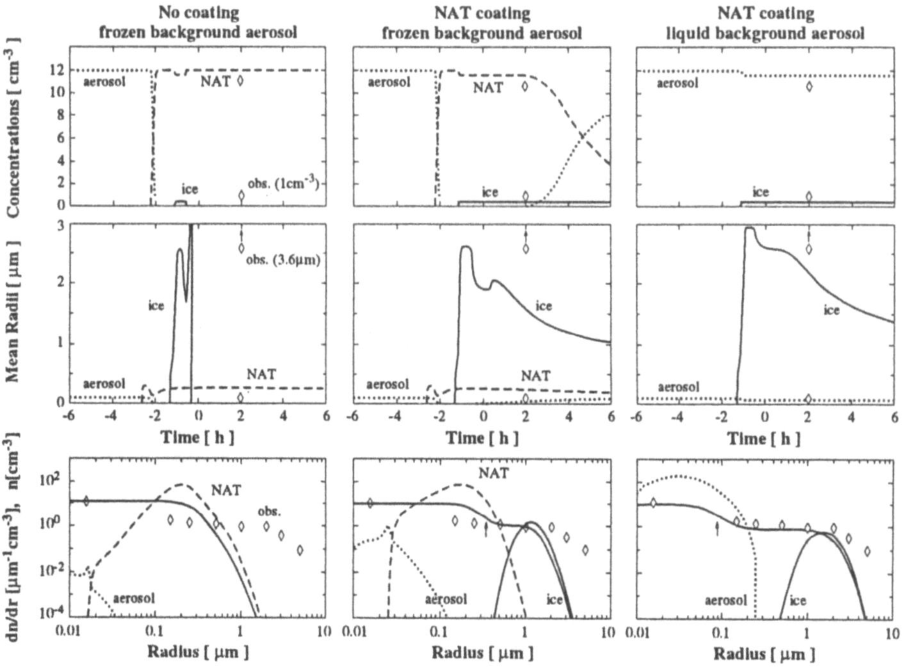

Figure 7: Model results for three specific cases: no coating (left), NAT-coating and solid background aerosol (center), NAT-coating and liquid background aerosol (right column). Upper row: particle number densities (weighted by $(T/T_0)(p_0/p)$); center row: mean radii; lower row: differential (thin) and integral (thick line) particle number densities at t=+2 h. Balloon-borne measurements are shown as diamonds for the thin particle layer at 21.54 km altitude (Deshler et al., 1994).

program simulates this by stopping ice evaporation). After t=+0.5 h NAT remains undersaturated and ice evaporation is assumed to proceed with β_{mix}. At the time of measurement at t=+2 h the mean radius is still $\bar{a}_{ice} = 1.5\mu m$. In disagreement with the measurement, the relatively large NAT particles push the shoulder (see arrow) of the bimodal distribution to large radii (Figure 7 bottom, center).

*Case*3: NAT-coating, liquid background aerosol. In this case NAT particles do not form so that the bulk of the HNO_3 remains in the gas phase and is fully available for co-condensation on ice (Figure 7 bottom, right). The reasonable agreement found in the region of the bimodal shoulder (see arrow) suggests that NAT particles were not present, their nucleation being prevented by the liquid state of the aerosol. At t=+2 h the mean ice radius is $\bar{a}_{ice} = 2\mu m$. The fact that this is still considerably lower than the observations (3.6 μm) might be related to our estimate of the supercooling with respect to ice: using $T < T_{ice} - 1.2K$ as condition for ice nucleation reproduces the observed radii, but ice particle number densities are then only 0.2 cm^{-3}.

3. CONCLUSIONS

Our knowledge of ice crystals in the atmosphere in general and in the cold regions of the upper troposphere and lower stratosphere in particular is still very limited. Moreover, there are very few laboratory measurements of g/e rates of ice crystals under conditions relevant to the atmosphere. This paper discusses several processes that may influence the growth and evaporation rates of ice crystals in the atmosphere, and consequently the lifetimes and radiative properties of cirrus. Of tropospheric importance are shape and surface impedance effects, both of which may increase the particle lifetimes at low supersaturations by factors up to 5 - 10 compared with predictions of lifetimes for spherical ice particles that ignore crystal surface effects.

In the lower polar winter stratosphere nitric acid hydrates appear to be more stable than water ice, Hence here the largest effect on ice particle lifetime may stem from the possibility of having high impurity concentrations in the ice or forming eutectic layers of hydrates and ice. Such coating mechanisms in principle could lead to a very substantial lengthening of ice particle lifetime. Preliminary results indicate that changes of 1--2 orders of magnitude could be expected, but when temperatures are around or above the equilibrium temperatures of the coating hydrates the theory is far from being mature or experimentally proven. More observations are needed to established a firm understanding of these mechanisms.

Acknowledgments: MBB was partially supported by NSF grant ATM 92-23563. We thank J. Nelson for providing us with new calculations from the pseudocoupled model and B. Swanson, for calculations of the effects of impurities on surface melting.

REFERENCES:

Arnott, W., Y. Dong, J. Hallett, 1994: Role of small ice crystals in radiative properties of cirrus: a case study. FIRE II. J. Geophys. Res. 99, 1371-1381.

Braham, R.R., 1967: Cirrus cloud seeding as a trigger for storm development, J. Atmos. Sci., 24, 311-312.

Braham, R.R. and P. Spyers-Duran, 1967: Survival of cirrus crystals in clear air, J. Appl. Meteor., 6, 1053-1061.

Burton, W., N. Cabrera and F. Frank, 1951: The growth of crystals and the equilibrium structure of their surfaces. Proc. Roy. Soc. (London) A243, 299.

Chen, J. and D. Lamb, 1994: The theoretical basis for the parameterization of ice crystal habits: growth by vapor deposition. J. Atmos. Sci. 51, 1206-1221.

Clapsaddle, C. and D. Lamb, 1989: The sorption behavior of SO2 on Ice at temperatures between -30 C and -5 C. Geophys. Res. Lett. 16, 1173-1176.

Conklin, M., R. Sommerfeld, S. Laird and J. Villinski, 1993: Sulfur dioxide reactions on ice surfaces: implications for dry deposition to snow. Atmos. Env. 27A, 159-166.

Dash, J. G., 1989: Surface melting. Contemporary Physics 30, 89-100.

Deshler, T., Th. Peter, R. Müller and P.J. Crutzen, 1994: The lifetime of leewave-induced ice particles: I. observations, Geophys. Res. Lett., 21, in press.

Dye J.E., D. Baumgardner, B.W. Gandrud, S.W. Kawa, K.K. Kelly, M. Loewenstein, G.V. Ferry, K.R. Chan and B.L. Gary, 1992: particle size distributions in Arctic polar stratospheric clouds, growth and freezing of sulfuric acid droplets, and implications for cloud formation, J. Geophys. Res., 97, 8015--8034.

Elbaum, M. and M. Schick, 1991: Application of the theory of dispersion forces to the surface melting of ice. Phys. Rev. Lett.66, #13,1713-1716.

Frank, F., 1982: Snow crystals. Contemp. Phys. 23, 3-22.

Goodman, J., O.B. Toon, R.F. Pueschel, K.G. Snetsinger, and S. Verma, 1989: Antarctic stratospheric ice crystals, J. Geophys. Res., 94, 16449-16457.

Hall, W.D. and H.R. Pruppacher, 1977: The survival of ice particles falling from cirrus clouds into subsaturated air, J. Atmos. Sci., 33, 1995-2006.

Hanson, D., and K. Mauersberger, 1988: Laboratory studies of nitric acid trihydrate: implications for Antarctic ozone depletion, Geophys. Res. Lett., 15, 855--858.

Heymsfield, A.J., 1973: Cirrus uncinus generating cells and the evolution of cirriform clouds. Ph.D. Thesis, Dept. Geophys. Sci., University of Chicago, Ill.

Heymsfield, A. and C. Platt, 1984: A parameterization of the particle size spectrum of ice clouds in terms of the ambient temperature and the ice water content, J. Atmos. Sci. 41, 846.

Kinne, S. and K. N. Liou, 1989: The effects of the nonsphericity and size distribution of ice crystals on the radiative properties of clouds. Atmos. Res. 24, 273-284.

Kuroda, T., 1984: Rate determining processes of growth of ice crystals from the vapour phase. J. Cryst. Gr. 12, 21.

Luo, B.P., Th. Peter, and P.J. Crutzen, 1994: Freezing of sulfuric acid aerosol droplets, Geophys. Res. Lett., 21, in press.

MacKenzie, A. and P. Haynes, 1992: The influence of surface kinetics on the growth of stratospheric ice crystals. J. Geophys. Res. 97, D8, 8057-8064.

Molina, M., 1992: The probable role of stratospheric 'ice' clouds: heterogeneous chemistry of the 'ozone hole". In CHEMRAWN VII: Chemistry of the Atmosphere: The Impact of Global Change. ed J. Calvert. Blackwell sci. Pub., Oxford.

Nakaya, U, M. Hanajima and J. Mugurama, 1958: Physical investigations of the growth of snow crystals. J. Faculty of Sci. Hokkaido Univ., Jpn, Ser. II, vol V, #3, 88.

Nelson, J., 1994: A theoretical study of ice crystal growth in the atmosphere. PhD dissertation, University of Washington

Nelson, J. and M. Baker, 1994: A new model for vapor growth of small ice crystals in the atmosphere. In Press.

Peter, Th., R. Müller, P. Crutzen and T. Deshler, 1994: The lifetime of leewave induced ice particles in the Arctic stratosphere: II. stabilization due to NAT coating. To be published, Geophys. Res. Lett. 21.

Pruppacher, H.R. and J.D. Klett,, 1978: Microphysics of Clouds and Precipitation, D. Reidel Publ. Co., Dordrecht.

Sassen, K., M. Griffin and G. Dodd, 1989: Optical scattering and microphysical properties of subvisual cirrus clouds, and climatic implications. J. Appl. Met. 28, 91.

Schlager, H., F. Arnold, D. Hofmann, and T. Deshler, 1990: Balloon observations of nitric acid aerosol formation in the Arctic stratosphere: gaseous nitric acid, Geophys. Res. Lett., 17, 1275--1278.

Solomon, S., R. Garcia, F. Rowland and D. Wuebbles, 1986: On the depletion of Antarctic ozone. Nature 321, 755-758.

Strom, J., J. Heintzenberg, K. J. Noone, K. B. Noone, J. Ogren, F. Albers and M. Quante, 1994: Small crystals in cirriform clouds: a case study of residue size distribution, cloud water content and related cloud properties. Atmos. Res. 32, 125-141.

Strom, J. and J. Heintzenberg 1994: Water vapor, condensed water and crystal concentration in orographically influenced cirrus clouds. To be published, J. Atmos. Sci.

Tolbert, M.A. and A.M. Middlebrook, 1990: Fourier transform infrared studies of model polar stratospheric cloud surfaces: growth and evaporation of ice and nitric acid/ice. J. Geophys. Res., 95, 22423--22431.

Toon, O.B., R.P. Turco, J. Jordan, J. Goodman and G. Ferry, 1989: Physical processes in polar stratospheric ice clouds. J. Geophys. Res., 94, 11359--11380.

Valdez, M. and G. Dawson, 1989: Sulfur dioxide incorporation into Ice depositing from the vapor. J. Geophys. Res. 94, 1095-1103.

Volkert, H. and D. Intes, 1992: Orographically forced stratospheric waves over northern Scandinavia, Geophys. Res. Lett., 19, 1205--1208.

Wofsy, S.C., R.J. Salawitch, J.H. Yatteau, M.B. McElroy, B.W. Gandrud, J.E. Dye and D. Baumgardner, 1990: Condensation of HNO3 on falling ice particles: mechanism for denitrification of the polar stratosphere, Geophys. Res. Lett., 17, 449--452.

WMO, "Scientific Assessment of Ozone Depletion: 1991", World Meteorological Organization, Global Ozone Research and Monitoring Project, Report No. 25, Geneva.

Abrupt climatic changes : A global perspective from ice cores

Jean JOUZEL[1,3], Claus HAMMER[2], Claude LORIUS[3], Sigfus JOHNSEN[2], Pieter GROOTES[5], Michel STIEVENARD[1] and Jim WHITE[6]

1) Laboratoire de Modélisation du Climat et de l'Environnement, CEA/DSM, CE Saclay 91191, Gif/Yvette Cedex France.
2) Niels Bohr Institute of Astronomy, Physics and Geophysics, Department of Geophysics, University of Copenhagen, Haraldgade 6, DK 2200, Copenhagen, Denmark
3) Laboratoire de Glaciologie et Géophysique de l'Environnement du CNRS, BP96, 38402, St Martind'Hères Cedex France.
4) Science Institute, University of Reykjavik, Dunhaga 3, Reykjavik 107, Iceland
5) Department of Geological and Quaternary Research center, University of Washington, Seattle, Washington, 98195, USA.
6) Institute of Arctic and Alpine Research Institute and Department of Geological Sciences, Campus Box 450, University of Colorado, Boulder, Colorado 80309, USA

On July 12 1992, the GRIP European deep drilling (GReenland Ice Core Project) reached the bedrock in Central Greenland (Summit region; altitude of the site above sea level, a.s.l., is 3250 m, Figure 1). One year later, the US Greenland Ice Sheet Project (GISP2) successfully completed a second drilling in the same region at a site located 28 km apart (3208 m a.s.l.). These two, more than 3 km long, cores have now been continuously analysed for their ice oxygen 18 content. Before reviewing the important climatic information that has already been derived from those profiles and putting them in a global perspective through a comparison with the Vostok deuterium record, we briefly summarize why and how these isotopic changes provide a record of the local temperature at the time of formation of the precipitation.

The saturation vapor pressures of HDO et $H_2{}^{18}O$, the most abundant isotopic forms of water (D is for deuterium), are slightly lower than that of

NATO ASI Series, Vol. I 35
Clouds, Chemistry and Climate
Edited by Paul J. Crutzen and Veerabhadran Ramanathan

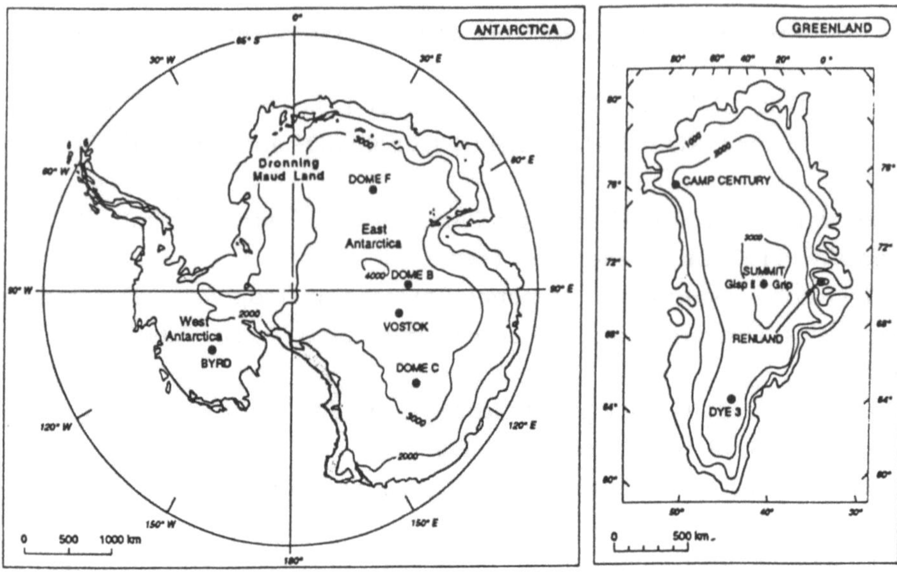

<u>Figure 1</u> : Map of Greenland and Antarctica with indication of deep drilling sites

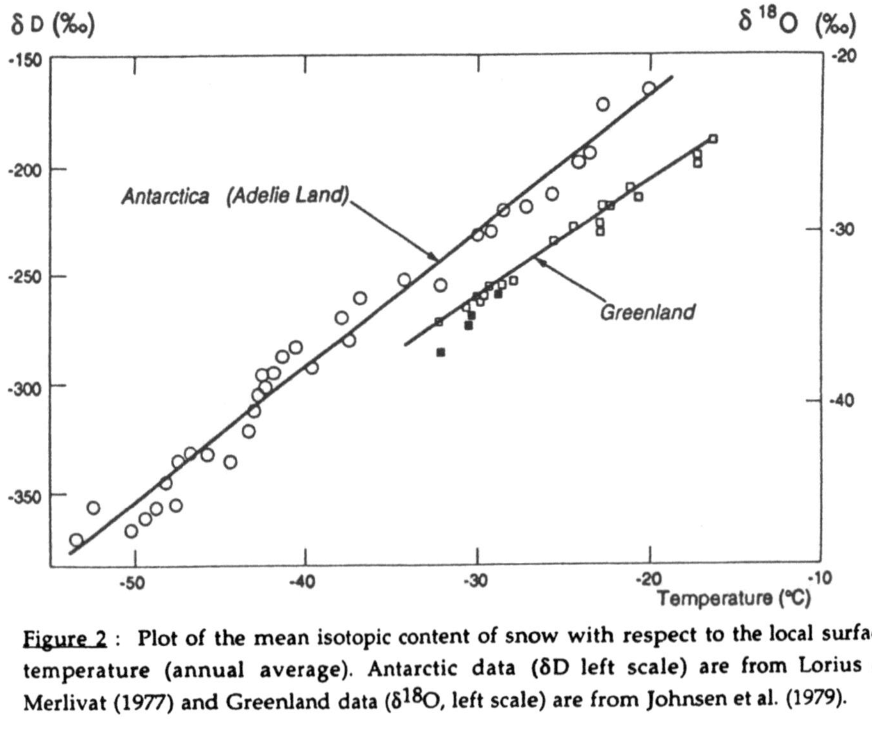

<u>Figure 2</u> : Plot of the mean isotopic content of snow with respect to the local surface temperature (annual average). Antarctic data (δD left scale) are from Lorius et Merlivat (1977) and Greenland data ($\delta^{18}O$, left scale) are from Johnsen et al. (1979).

$H_2{}^{16}O$. As a result, the condensed phase (rain or snow) is isotopically enriched with respect to the vapor phase and there is a decrease of the isotopic contents (both for vapor and condensed phases) as a given airmass is progressively depleted by the formation of precipitation (Dansgaard, 1964). This is why deuterium and oxygen 18 precipitation contents (δD and $\delta^{18}O$)* generally decrease from equatorial to polar regions. In Greenland (Johnsen et al., 1989) and in Antarctica (Lorius and Merlivat, 1977), this results in a linear relationship between the mean isotopic content of snow (δD or $\delta^{18}O$) and the mean annual temperature (Figure 2). These empirical isotope/temperature relationships, satisfactorily explained through simple Rayleigh type isotopic models (Dansgaard, 1964; Jouzel and Merlivat, 1984) serve as a basis for reconstructing past temperatures from ice core using either δD or $\delta^{18}O$ isotopic profiles. Figure 2 shows that a cooling of 1°C (surface temperature) corresponds to a decrease of .67‰ in oxygen 18 in Greenland, Johnsen et al., 1989) and of 6‰ in deuterium in Antarctica (Lorius and Merlivat, 1977) but due to the existence of a strong inversion this value is of 9 ‰ when taken with respect to the atmospheric temperature.

In this approach, glaciologists assume that the temporal isotope/temperature slopes (in a given site for different climatic periods) is similar to the spatial slopes defined for the current climate. The validity of this critical assumption may be examined by incorporating water isotope cycles and associated fractionation processes in General Circulation Models of the atmosphere (Joussaume et al., 1984; Jouzel et al., 1987). The advantage is twofold : the complexity of atmospheric processes is then taken into account and the distribution of isotopes in precipitation may be simulated for different climates, modern and glacial for example. Such a comparison has now been performed by two modeling groups (Laboratoire de Météorologie Dynamique in CNRS/PARIS and Goddard Institute for Space Studies in NASA New-York) leading to the same overall conclusion : spatial and temporal slopes are quite similar in polar regions (Joussaume and Jouzel, 1993; Jouzel et al., 1994). The

---

* D/H et $^{18}O/^{16}O$ ratios are commonly expressed in δ notation with respect to V-SMOW (Vienna Standard Mean Ocean Water) with isotopic ratios R_{SMOW} of D/H = $155.76 * 10^{-6}$ and of $^{18}O/^{16}O = 2005.2 * 10^{-6}$. δ values are given by : $\delta = (R_S / R_{SMOW})) - 1$, in which R_S is the isotopic ratio of the sample, and expressed in per mill.

use of the spatial slope, the only one we have access to, is thus justified for reconstructing temperature changes in the past. From those model simulations, the uncertainty on the temperature estimate appears to be of the order of 30%. Part of this uncertainty could be due, at least for Greenland, to glacial-interglacial changes in the origin of the air masses which provide precipitation (Charles et al., 1994).

Analysing both deuterium and oxygen 18 may provide information about such changes. The approach is based on the existence of a second isotopic effect due to the fact that molecular diffusivities in air are lower for isotopic molecules than for $H_2{}^{16}O$. This induces an additional effect for non-equilibrium processes, evaporation and condensation. The isotopic effect resulting from the difference in vapor pressures is 8 to 10 times higher for deuterium than for oxygen 18 whereas this additional effect is of the same order in both cases. This difference implies that oxygen 18 is more sensitive than deuterium to non-equilibrium processes and it is the reason why, in general, it is interesting to analyze both isotopic species. Interpretation is currently done through the deuterium-excess parameter defined as $d = \delta D - 8*\delta^{18}O$. The application of Rayleigh models (Merlivat and Jouzel, 1979; Johnsen et al., 1989) shows that the deuterium/oxygen 18 relationship largely depends on the temperature and humidity in the source region of precipitation. The deuterium-excess parameter thus tells us about the conditions prevailing at the origin of precipitation which very usefully complement those about the temperature of the site derived either from deuterium or from oxygen 18 variations.

GRIP et GISP 2 : A series of abrupt changes

The originality of the GRIP and GISP 2 cores holds, for a large part, in that they are the first Greenland records going back beyond the last glacial period. Questions raised by the apparent instability during the previous interglacial (Eemian) and by the divergence between the two cores for their lower part presently retain much of the attention of the scientific community. However very remarkable and new information has been derived from these two cores both for the last deglaciation and for the last climatic period.

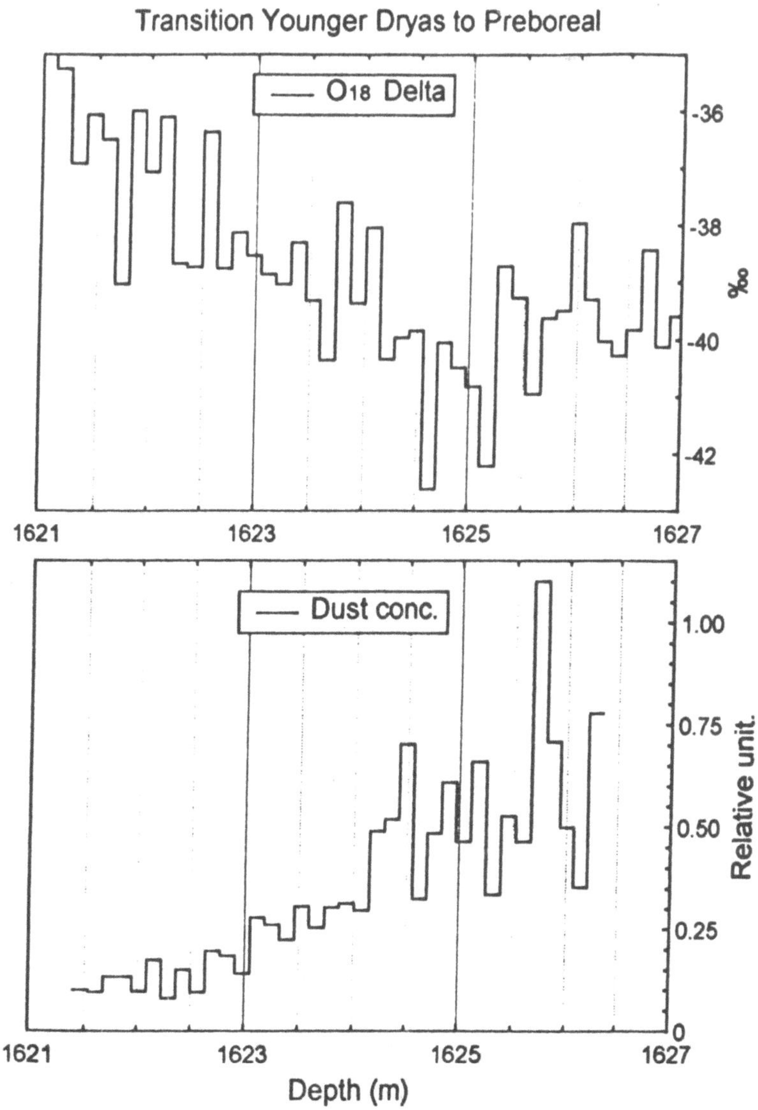

Figure 3 : Transition from Younger Dryas to Pre Boreal in the GRIP core. The major change in the dust concentration took place in 20-25 years, while the change in the oxygen 18 content lasted approximately 50 years. The dust concentration is given in relative units, a unit corresponding approximately to 1.3 mg/g of ice.

The last deglaciation is a well documented period characterized by a first warming trend which culminates during the Bölling and is followed by a return to colder conditions (Younger-Dryas), particularly well marked in the North atlantic and adjacent continental regions (Rind et al., 1986). GRIP (Johnsen et al., 1992) and GISP 2 records (Grootes et al., 1993) confirm the rapidity of the transition towards the current Holocene climate. New information concern the chronology of these events and the discovery that beyond temperature, even more rapid changes affect the precipitation rate and the atmospheric circulation.

The existence of seasonal variations (dust, chemical elements) allow precise absolute dating of the deglaciation. In agreement with the recent ^{14}C recalibration (Bard et al., 1993), it is confirmed that the Younger-Dryas/Holocene (Pre Boreal) transition is older than previously assumed. In Greenland, the Younger Dryas ended around 11.6 kyr BP (11600 years ago i.e. 11550± 70 at GRIP, Johnsen et al., 1992; and 11640± 250 at GISP 2, Alley et al., 1993).

Figure 3 illustrates how rapidly ended this transition in the GRIP core. It took less 20-25 years for the wind-blown continental dust content of the atmosphere to change from high glacial to much lower interglacial values while the change in the isotopic values, which is estimated to correspond to a warming of 7°C in Greenland, lasted approximately 50 years. This finding confirms the results obtained on the Dye 3 core in which deuterium-excess (not yet measured in the GRIP core at such a detailed resolution) also shows a change as rapid as that underwent by dust (Dansgaard et al., 1989). This reflects rapid reorganizations in atmospheric circulation and rapid retreat of the North Atlantic sea ice cover with associated change in the origin of Greenland precipitation.

This rapid changes in atmospheric circulation is fully confirmed from the GISP2 core analysis (Taylor et al., 1993) through the continuous electrical conductivity profile which indirectly depends on the dust fallout as ice formed in dusty periods has a high concentration of alkaline dust which significantly reduces the conductivity compared to less dusty periods. The snow accumulation doubled even more rapidly, possibly in one to three years; the accumulation increase at the onset of the Bölling was also large and abrupt (Alley et al., 1993). Unlike the isotopic record which gives access to a local

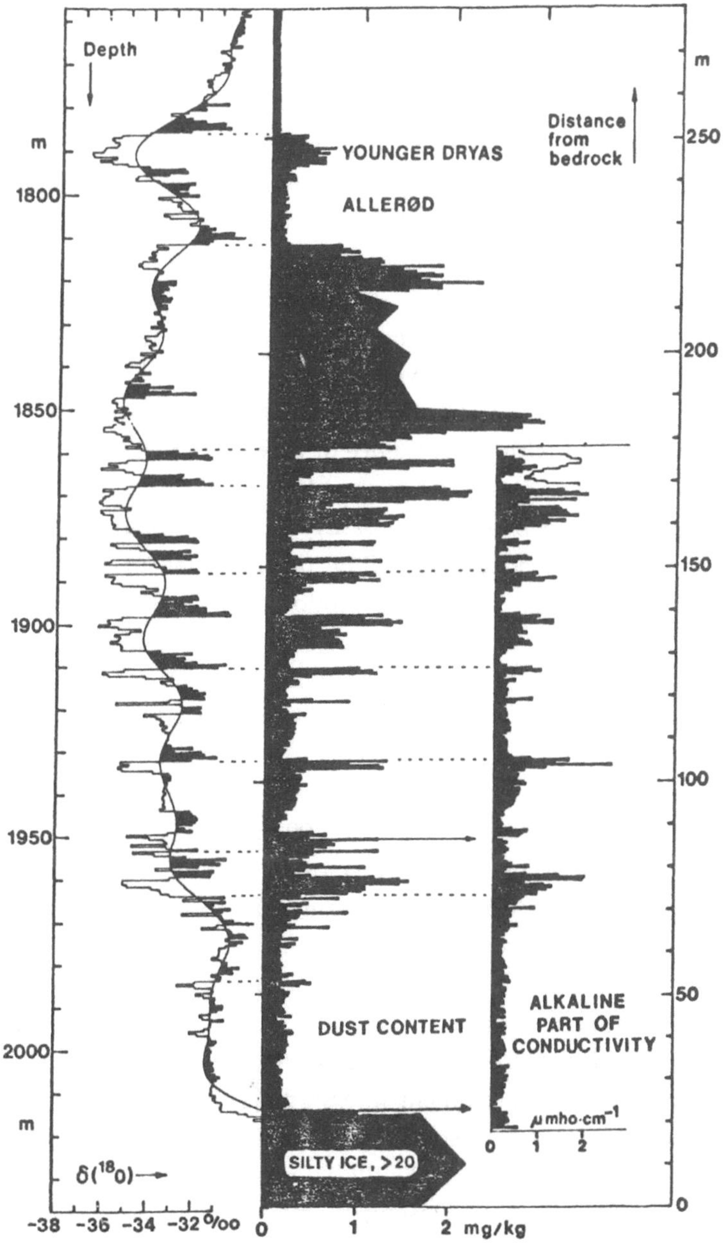

Figure 4 : Ice age part of the Dye 3 deep core showing δ18O, dust concenntrations, and alkalinity in 55 cm average samples. From 1821 to 1852 m only selected segments were measured for dust, which gives the dust profile a sawtooth appearance in this depth

parameter (temperature) these rapid changes particularly those in the accumulation rate are indicative of modifications occuring on a larger geographical extent.

Figure 4 shows the Dye 3 dust profile from the Younger-Dryas downward along with the alkaline part of conductivity, which as mentioned above directly depends on the dust fallout, and the oxygen 18 climatic profile (Hammer et al., 1985). During the glacial period there are rapid changes of both climate and dust content. Closer examination of these events (Dansgaard et al., 1984) shows that both parameters are anticorrelated with high dust content for colder periods and vice-versa.

For the last glacial period, the most important results of the central Greenland records deal with the existence of these rapid climatic changes (now so-called the "Dansgaard-Oeschger" events) discovered and well characterized in the Camp Century and Dye 3 cores (Dansgaard et al., 1984). Rapid warmings, often more than half of that corresponding to the glacial-interglacial difference and taking place in a few decades, are followed by a slower cooling and a generally rapid return to glacial conditions. The existence and characteristics of these events are fully confirmed : about 20 of such interstadials lasting between 500 and 2000 years occured during the last glacial period (Dansgaard et al., 1993, Grootes et al., 1993). Rapid changes in dust concentrations, with higher values during the colder periods, are systematically associated with these interstadials (GRIP project members, 1993).

The general shape of these interstadials is quite similar to the rapid changes recently discovered in North Atlantic deep-sea core records and the most prominent of these interstadials may be associated with the sedimentary "Heinrich" layers interpreted as reflecting massive iceberg discharge from Northern Hemisphere ice sheets (Bond et al., 1993). Associated temperature warmings recorded in Greenland and in the North Atlantic are quite comparable and the most prominent interstadials may be easily identified in the marine records (Figure 5) despite slight age differences attributable that the marine and ice records are independently dated. There is thus quite probably a link between the "Dansgaard-Oeschger" events and ice cap instabilities through associated changes in the oceanic circulation.

A significant increase in methane concentration is associated with these warm interstadials, suggesting that they were at least hemispheric in their extent (Chappellaz et al., 1993). These authors proposed that variations in the hydrological cycle at low latitudes may be responsible for the variations in both methane and Greenland temperature during these interstadials (through changes in this hydrological cycle over, respectively the continent and the ocean).

Beyond 40 kyr BP, a preliminary chronology of the GRIP core (Figures 6 et 9) was derived using a glaciological model and assuming that the 110 kyr BP level is correctly identified through a comparison with the marine record [Dansgaard et al., 1993). The depth interval between 2775 and 2876 m corresponds to the 140/110 kyr BP period that fully encompasses the last interglacial (Eemian) considered, in general, as being warmer than the current Holocene. This appeared to be partly confirmed (Figure 6) but the variability of the isotopic record was totally unexpected (GRIP project members, 1993) : periods warmer than the present climate (up to ~ 4°C) are interrupted by a series of high amplitude oscillations with isotopic values plunging to mid-glacial levels. These cold events began extremely rapidly (a few decades) and lasted from 70 years for the shortest to 7000 years for the longest; the associated temperature changes are of the order of 10°C. GRIP scientists noted the difficulty to interpret these rapid changes in the absence of the large Laurentide and Fennoscandian ice-sheets which constitute potential sources of instabilities.

Questions about the validity of interpreting those isotopic changes in terms of climatic changes arise because there are significant differences between the GRIP and the GISP 2 records for the Eemian (Grootes et al., 1993). Whereas the two cores are in excellent agreement back to the level corresponding to 87 kyr BP in the GRIP chronology, the series start to diverge beyond (Figure 6). This divergence shows that ice flow must have altered the chronological sequences of the stratigraphy for the bottom part of one or both of the cores. Inclined layers generally associated with distortions due to folding (Boulton, 1993) are not observed at GRIP during a large part of the Eemian (as illustrated in Figure 6 they are observed only below 2847 m a level corresponding to 129 kyr BP) whereas they are observed at GISP2 as soon as the records begin to diverge (2678 m). GRIP is located on the Dome unlike GISP2 which might explain this different behaviour but the argument is weak at

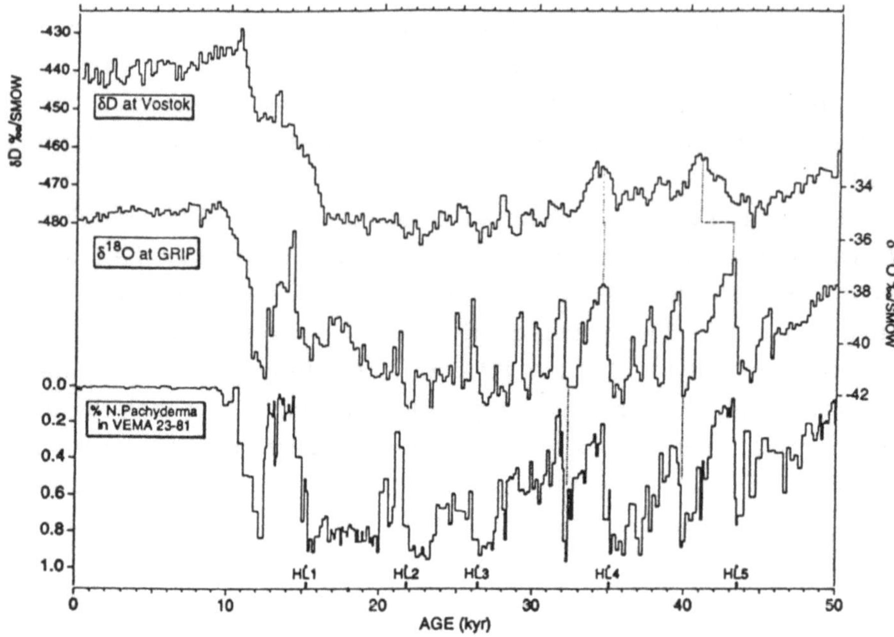

Figure 5 : Comparison, over the last 50 ka, of the climatic records 1) from Vostok (deutérium content), 2) from GRIP (oxygen 18 content) and 3) from the North Atlantic site V23-81 (percentage of N.Pachyderma, a proxy of sea-surface temperature) for which we have indicated the levels of the Heinrich layers (adapted from Bond et al., 1993).

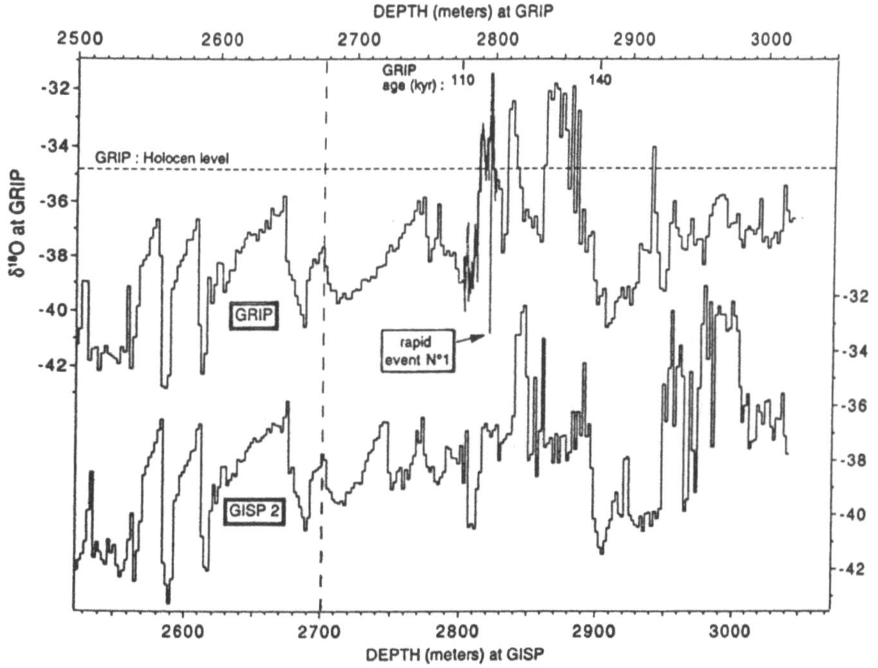

Figure 6 : Comparison of the GRIP and GISP2 isotopic records for the bottom part of the cores with respect to depth. Down to ~ 2700 m the agreement between the two records is excellent (the depth scales have been slightly shifted to fully illustrate this agreement). Over the GRIP profile corresponding to a 2.2 m step (2 m for GRIP), we have superimposed, around 2790 m, the detailed record of rapid event 1 (GRIP project members, 1993). We have indicated the GRIP levels coreesponding to 110 and 140 ka (Eemian) in the GRIP preliminary timescale and the level corresponding to the mean isotopic value during the Holocene. Hatched areas correspond to the depths for which inclined layers are observed in each of the cores.

least as long as the influence of possible movements of the dome over the last climatic cycle, that may not be ruled out, has been fully assessed.

The analysis of air bubble composition offers a promising approach to identify how, quite certainly for GISP2, possibly for GRIP, the Eemian and older ice has been distorted. The idea is to compare these atmospheric composition records which have a global significance with those obtained along the Vostok ice core (Barnola et al., 1987; Chappellaz et al., 1990; Sowers et al., 1993) where ice formed during the last interglacial, now far from the bedrock (~ 2 km) provides undisturbed ice sequences. Unlike air composition, the comparison of Greenland and Antarctic isotopic profiles discussed below is not a way to shed light on the GRIP/GISP2 divergence. Rather, its first aim is to examine similarities and differences between climate of both polar regions and then to better assess the character either regional, or global of the climatic events recorded in central Greenland.

Antarctica and Greenland : similarities and differences

Our knowledge of the relationship between climate and concentrations in greenhouse gases at the glacial-interglacial timescale (Figure 7) is largely based on the results obtained along the ice cores drilled at the Russian station of Vostok since the beginning of the seventies. French teams are associated with the program since about ten years and American teams joined it more recently. As illustrated in figure 7 where are reported the temperature record derived from the deuterium profile and the composition of air bubbles in CO_2 and CH_4 the records now cover the last ~ 220 ka (Jouzel et al., 1993). Beyond these records which contain information on climate sensitivity useful with respect to the future evolution of our climate (Lorius et al., 1990), the variations of many other parameters have been measured in the Vostok ice. They concern, in particular, the concentrations in various aerosols and chemical species and in cosmogenic or other isotopes.

We have selected in figure 8 some parameters which are of particular relevance to the relation between climate, radiative forcing and chemistry, the main focus of the series of articles presented in this book. Curves b and d relate respectively to the marine (sodium) and terrestrial (calcium) aerosol. In both cases, the inputs were higher during cold climatic conditions than during

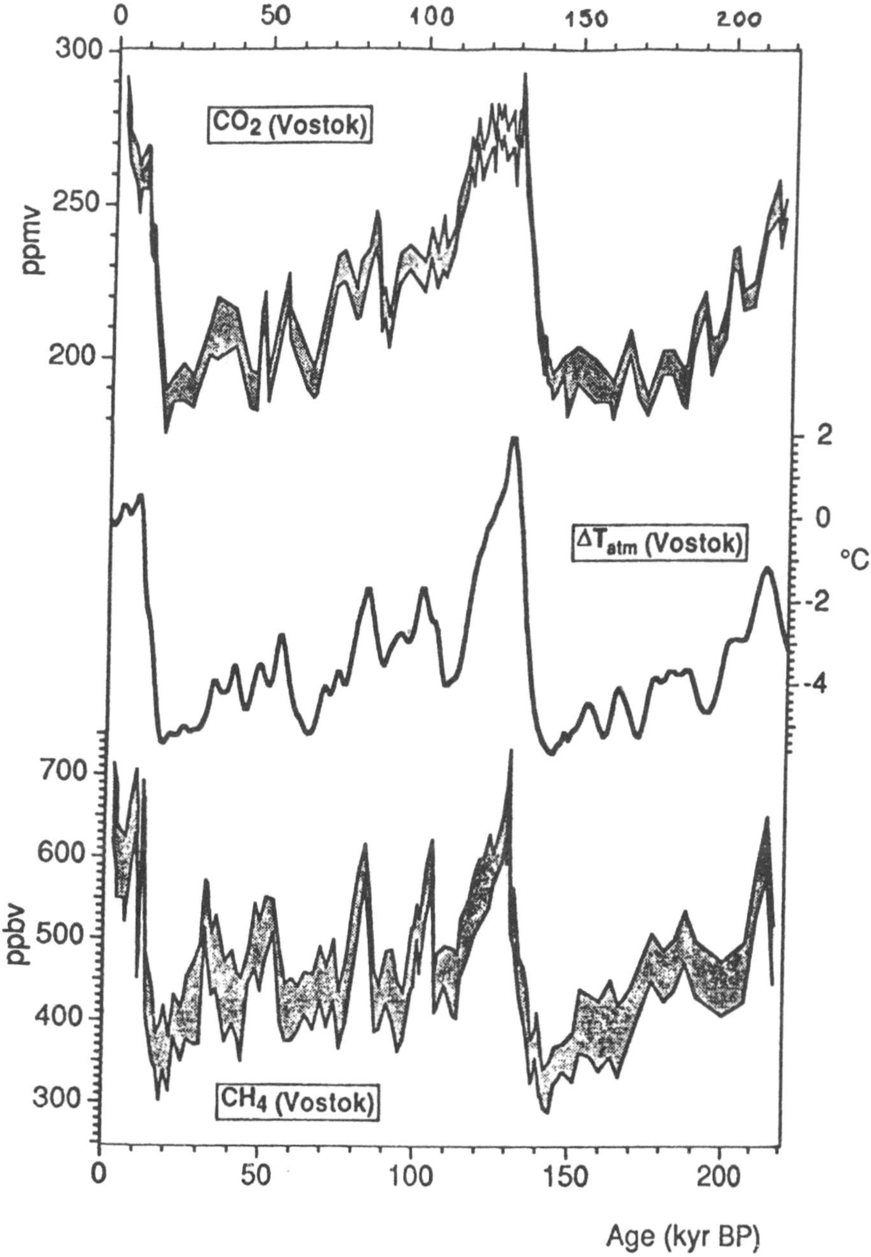

<u>Figure 7</u> : CO_2, CH_4 and atmospheric temperature (derived from the deuterium profile) with respect to time along the Vostok core over the last 220 ka . For CO_2 and CH_4, the envelope corresponds to the measurement accuracy (adapted from Jouzel et al., 1993).

interglacial periods. High concentration peaks are in particular observed during the very cold climate conditions characterizing the Last Glacial Maximum and the end of the penultimate glacial period. These peaks reflect strengthened sources and transport during full glacial conditions linked to higher windspeed, more extensive areas on the continents and the greater exposure of continental shelves; curve c (chlorine/sodium ratio) simply illustrates that the composition of the marine aerosol may be affected during its transport to Central Antarctica through fractionation processes themselves climate dependent (Legrand et al., 1988). The climatic impact of the change in atmospheric optical depth due to these variations in aerosol loading has been examined through various approaches (Harvey 1988, Lorius et al., 1990, Anderson and Charlson, 1990) but no definite conclusion has yet been reached. On the other hand, the sulfuric acid Vostok record (not shown) remains relatively stable over the whole climatic cycle which would indicate the absence of a long-term relationship between volcanism and climate.

The last two records, Methane Sulfonate acid (e) and non-seasalt sulfate (f), two chemical compouds derived from dimethylsulfide emitted from the marine biota, deal with the problem of the relationship between marine phytoplancton, atmospheric sulfur cycle, cloud albedo and climate (Charlson et al., 1987, Schwartz, 1988). In assuming that the global average number population of cloud condensation nuclei is proportional to non- seasalt sulfate Legrand et al. (1988) suggested that the resulting change in cloud albedo would have participated for about 1°C to the cooling of the Last Glacial Maximum. This evaluation has to be further assessed through complementary experimental and modeling studies but one result that, in any event, clearly comes from such ice core profiles is the extreme sensitivity of the ocean-atmosphere sulphur cycle to climate change (Legrand et al., 1991). Although such a sulfur cycle/climate link is also seen in Greenland, its nature may be different in each hemisphere as revealed by a study of MSA and sulfate performed along the Renland ice core (Figure 1) showing decreasing concentration of MSA during cold climate but increasing concentrations of non-seasalt sulphate (Hansson and Saltzman, 1993).

Coming now to the main objective of the present article which is to put the abrupt climatic changes recorded in the Greenland ice in a global

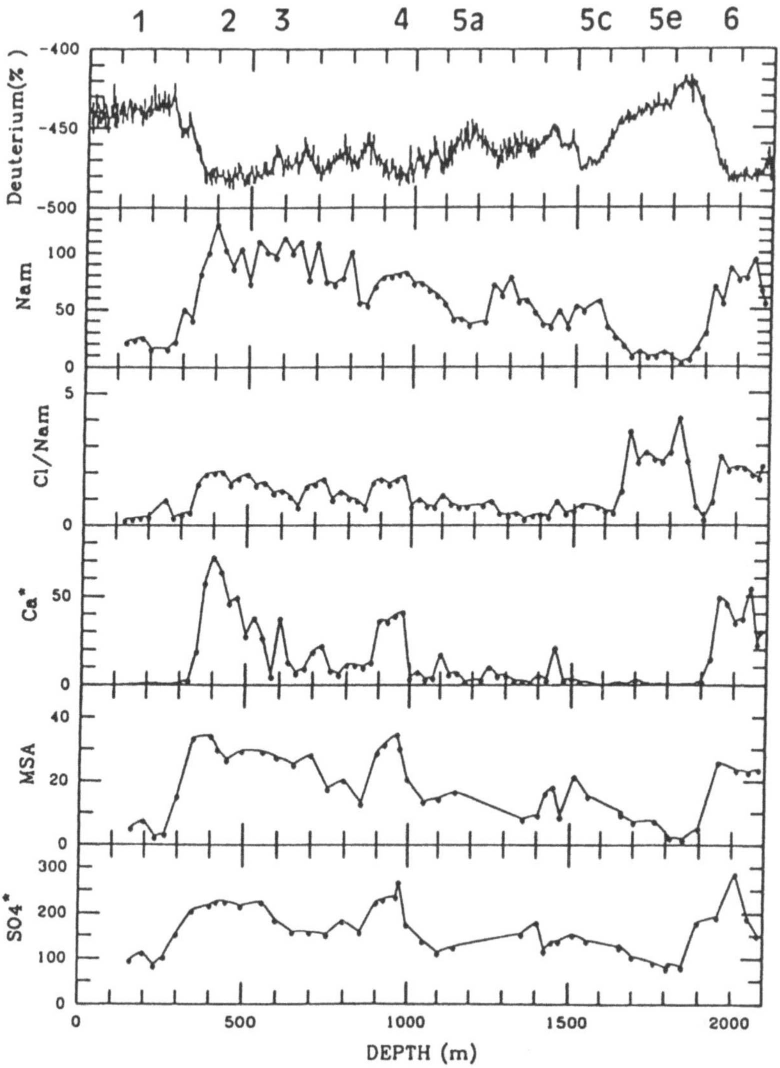

Figure 8 : Vostok ice core profiles with respect to depth over the last glacial-interglacial cycle (the last 150 ka) : a) deuterium profile, b) marine sodium, c) chlorine/marine sodium weight ratio, d) excess-calcium (terrestrial), e) Methane Sulfonate Acid (MSA) and f) non-seasalt sulfate (adapted from Legrand, in press).

perspective, we now go through a comparison of the Vostok the Greenland core climatic records over the three periods we have discussed above.

Although the precision of the Vostok timescale established using a glaciological model is no better than ± 10% (Jouzel et al., 1993), some interesting results may be derived about the shape and timing of the Vostok deglaciation. This transition is also a two-step process (Figure 5) but the temporary cooling is about three times weaker than that observed in Greenland during the Younger-Dryas. This two-step feature has thus a large geographical extent as already showed from numerous oceanic and continental records essentially from the Northern Hemisphere but also from the southern Ocean (Labracherie et al., 1989). However the main deglaciation step started earlier at Vostok (~ 17 kyr BP) than in Greenland (~ 15 kyr BP) which also applies for the second warming period as previously showed for the Southern Ocean (Labracherie et al., 1989). This is as much true as the Vostok timescale of Figure 5 is probably too young (by more than 1000 years) for this part of the deglaciation (Chappellaz et al., 1993, Sowers et al., 1993). Also, warmings are much slower in Antarctica than in Greenland. Although oceanic deep circulation changes, with reduced formation of North Atlantic deep water during the LGM plays probably a key role in linking the Northern Hemisphere and Southern Hemisphere climates, these results illustrate the complex nature of this interhemispheric link. It raises the possibility that such oceanic circulation changes may have their origin in the Southern Hemisphere.

Another noticeable feature concern the fact that the Vostok record is characterized by an early Holocene temperature optimum (~ 11 kyr BP) which differs from what is observed is the Northern Hemisphere where the climatic optimum occured around 6 kyr BP. This difference probably due to the presence of the extensive remaining ice sheets in the Northern hemisphere (Ciais et al., 1992).

Those characteristics, two-step deglaciation and early climatic optimum, are common to the three Antarctic cores, Vostok, Dôme C, Dôme B that cover the deglaciation (Jouzel et al., 1995). Although, dating uncertainties prevent a reliable comparison with the Byrd record (Hammer et al., 1994), we note that the situation appears different in this West Antarctic site where there is no such a clear cooling possibly because this site is in the Pacific sector weakly

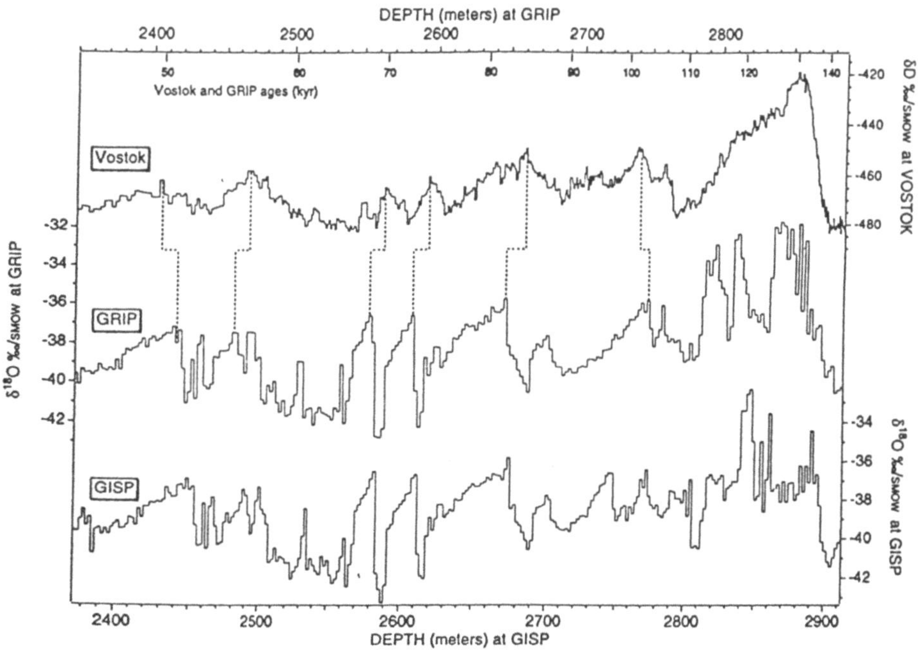

Figure 9 : Comparaison between the Vostok (deuterium content) and GRIP and GISP2 (oxygen 18 content) for a part of the glacial period and the Eemian. GRIP and GISP2 are reported with respect to depth (linear scales sligthly shifted for fully illustrating the agreement between the two records down to ~2700 m). We have indicated (top scale) the GRIP timescale (preliminary timescale from Dansgaard et al., 1993). The Vostok record is also reported with respect to age. We have indicated the correspondences we suggest between Vostok and GRIP for this part of the records.

influenced by oceanic circulation changes taking their source in the Atlantic. In any case, results obtained at Vostok and in the other sites of East Antarctica illustrate the quite global character of the two-step shape of the last deglaciation.

The observation that important North Atlantic climatic events have an attenuated counterpart in the Vostok record made for the last deglaciation seems to apply for the "Dansgaard-Oeschger" events. The most prominent GRIP interstadials may be identified in the Vostok record (Figures 5 and 9) but obviously, we have not yet any indication that the events connected by dotted lines are in or out of phase. Comparison of the smoothed GRIP and Vostok isotopic profiles shows similarities in the low frequency part of the records; the clear presence of the obliquity cycle (41 ka) in both records suggests that those similarities result from a common response to astronomical forcing (Dansgaard et al., 1993) the Greenland record being, in addition, strongly affected by non-linearities related to Northern Hemisphere ice sheet instabilities (Bond at al., 1993; Oerlemans, 1993).

As for the last deglaciation, the existence of an interhemispheric link through oceanic circulation changes may be reasonably proposed. However, one must be cautious about the visual correlations proposed in figures 5 and 9 and about the suggested mechanisms as well, as long as the validity of the chronologies based on glaciological models is not better established. Indeed, as an alternative mechanism, Yiou (1994) suggests that the Vostok record may be itself affected by instabilities of the Antarctic ice sheet which would be independent from those occuring in the Northern hemisphere. As for other periods, a detailed comparison of atmospheric composition changes offers a way to correlate the Central Greenland and Vostok ice cores independently from their isotopic records and thus to better examine what are the links between Greenland and Antarctic climates during the last glacial. The existence of well defined ^{10}Be peaks during the glacial period (Raisbeck et al., 1987) may also serve to correlate Antarctic and Greenland ice records over this period.

For the last interglacial, we focus here the discussion on GRIP and Vostok, the GISP2 record having, in any event for its Eemian part, a large probability to be seriously affected by ice flow distortion. GRIP and Vostok have interesting similarities : 1) the warmest part of the Eemian is in both records warmest than the Holocene (up to 4°C at GRIP, slightly less at Vostok); 2) in each case the Eemian lasts longer than the interglacial climatic stage 5e defined in the oceanic records. It is well established that in Antarctica, the temperature starts to rise at the end of the previous glacial well before global ice volume starts to decrease (Pichon et al., 1992).

Beyond these common features, the absence of rapid changes at Vostok both during the deglaciation that takes place in one single step and during the interglacial itself, constitutes a major difference between the GRIP and Vostok Eemian. Such a difference does not result from the difference in the time resolution of the records as illustrated in figure 6 where they are both showed with a 200 yr time step. Indeed, despite a ten times lower accumulation, the interglacial is represented by a larger depth interval at Vostok than at GRIP where the thinning is very important owing to the proximity of the bedrock (~ 200 m) : the time span -110, -140 kyr BP corresponds to a thickness of 420 m at Vostok instead of ~ 100 m at GRIP.

There is no doubt that this difference in structure adds to the questions raised by the GRIP/GISP2 comparison. The assumption that rapid Eemian changes may had not be felt in Antarctica is however still plausible as much as this site not situated in the atlantic sector is potentially less influenced by any change occuring in the Atlantic. But as we just mentioned there are similarities between GRIP and Vostok for the last deglaciation and we must understand why this not hold true for the previous transition ? Part of the explanation may lie in the fact that interhemispheric ocean circulation being a slow process, only the GRIP events lasting sufficiently long may be recorded in Antarctica.

We have now undertaken a high resolution study of the last interglacial at Vostok with the aim of getting a time resolution comparable (~10 years) with that which allowed to detect a high amplitude cold excursion estimated to have lasted ~ 70 years around 115 kyr BP in the GRIP record (event 1 in figure 6). The first results concern the transition from the penultimate glacial period to the interglacial. They confirm what is seen in the coarser record (see curve a of figure 8) i.e. the absence of the abrupt and important changes recorded in the GRIP record during the corresponding period. There is no data available yet for the period around 115 kyr BP.

Conclusion

Studies of the GRIP and GISP2 ice cores demonstrate that in contrast with the extreme stability of the Holocene period, large and rapid climatic changes occured during the last ice age and during the transition towards the present Holocene period. Those changes may occur on the timescale of a human life or less, at least in the North Atlantic where they are best documented. They affect a large variety of climate related variables : atmospheric temperature and circulation, precipitation patterns and hydrological cycle, temperature and circulation of the ocean.

The most prominent of these instabilities, last deglaciation cooling and major interstadials, seem to have attenuated counterparts in the Antarctic record of Vostok but further studies are clearly necessary to decipher the exact link between Greenland and Antarctic climates, and more generally between Northern Hemisphere and Southern Hemisphere climates, during these periods of general instability.

The GRIP Eemian record has also been interpreted in terms of climate instabilities that also may have occured at the timescale of a human life. The differences between GRIP and the two other records are not sufficient by themselves for disregarding this Eemian current interpretation of GRIP to the extent that a reasonable explanation could be provided (i.e, possibility of a different influence of flow deformation between GRIP and GISP2, and possibility of a different climate behaviour between GRIP and Vostok). However, the existence of rapid climatic variations to be confirmed requires that a way for assessing the respective contribution of climate and ice flow in the Greenland Eemian signals be found. In any case, identifying those possible rapid events in other ice, oceanic or continental records is indispensable for confirming that they really occured and for evaluating their geographical significance. In this search, we must however remember that Greenland and, for example, European temperatures do not necessarily have to follow the same pattern.

Acknowledgements : We are very indebted to all participants to the GRIP, GISP2 and Vostok projects for their cooperative efforts. We also thank agencies which either organized or financially supported these three major drilling programs namely, the European Science Fondation, the Commission of European Communities, the Carlsberg Fondation and the National Science Fondation in Belgium, Denmark, France, Germany, Iceland, Italy, Switzerland and United Kingdom for GRIP; the US National Science fondation for GISP2, the Russian Ministry of Sciences, the Programme National d'études de la Dynamique du Climat, the Fondation de France, the Commission of European Communities and the National Science Fondation for Vostok. We thank Laurent Labeyrie and Gerry Bond for providing V23-81 data and Michel Legrand for providing figure 8.

References

R. B. Alley, D. A. Meese, C. A. Shuman, A. J. Gow, K. C. Taylor, P. M. Grootes, J. W. C. White, M. Ram, E. D. Waddington, P. A. Mayewski and G. A. Zielinski (1993) Abrupt increase in Greenland snow accumulation at the end of the Younger Dryas event, Nature, 362: 527-529.

T. L. Anderson and R. J. Charlson (1990) Ice-age dust and sea salt, Nature, 345: 393.

E. Bard, M. Arnold, R. G. Fairbanks and B. Hamelin (1993) $^{230}Th/^{234}U$ and ^{14}C ages obtained by mass spectrometry on corals, Radiocarbon, 35: 191-199.

J. M. Barnola, D. Raynaud, Y. S. Korotkevich and C. Lorius (1987) Vostok ice core provides 160,000-year record of atmospheric CO_2, Nature, 329: 408-414.

G. Bond, W. S. Broecker, S. J. Johnsen, J. Mc Manus, L. D. Labeyrie, J. Jouzel and G. Bonani (1993) Correlations between climate records from North Atlantic sediments and Greenland ice., Nature, 365: 143-147.

G. Boulton (1993) Two cores are better than one, Nature, 366: 507 -0508.

J. Chappellaz, J. M. Barnola, D. Raynaud, Y. S. Korotkevich and C. Lorius (1990) Ice core record of atmospheric methane over the past 160,000 years, Nature, 345: 127- 131.

J. Chappellaz, T. Blunier, D. Raynaud, J. M. Barnola, J. Schwander and B. Stauffer (1993) Synchronous changes in atmospheric CH_4 and Greenland climate between 40 and 8 kyr BP, Nature, 366: 443-445.

C. Charles, D. Rind, J. Jouzel, R. Koster and R. Fairbanks (1994) Glacial interglacial changes in moisture sources for Greenland : influences on the ice core record of climate, Science, 261: 508 - 511.

R. J. Charlson, J. E. Lovelock, M. O. Andrae and S. J. Warren (1987) Oceanic phytoplancton, atmospheric sulphur, cloud albedo and climate, Nature, 326: 655 - 661.

P. Ciais, J. R. Petit, J. Jouzel, C. Lorius, N. I. Barkov, V. Lipenkov and V. Nicolaïev (1992) Evidence for an early Holocene climatic optimum in the Antarctic deep ice core record, Climate Dynamics, 6: 169-177.

W. Dansgaard (1964) Stable Isotopes in Precipitation, Tellus, 16: 436-468.

W. Dansgaard, S. Johnsen, H. B. Clausen, D. Dahl-Jensen, N. Gundestrup, C. U. Hammer and H. Oeschger (1984) North Atlantic climatic oscillations revealed by deep Greenland ice cores, in Climate processes and climate sensitivity, edited by J. E. Hansen and T. Takahashi, 288 - 298, Am Geophys Union, Washington.

W. Dansgaard, S. J. Johnsen, H. B. Clausen, D. Dahl-Jensen, N. S. Gunderstrup, C. U. Hammer, J. P. Steffensen, A. Sveinbjörnsdottir, J. Jouzel and G. Bond (1993) Evidence for general instability of past climate from a 250-kyr ice-core record., Nature, 364: 218-220.

W. Dansgaard, J. W. White and S. J. Johnsen (1989) The abrupt termination of the Younger Dryas climate event, Nature, 339: 532-534.

GRIP project members (1993) Climatic instability during the last interglacial period revealed in the Greenland summit ice - core., Nature, 364: 203-207.

P. M. Grootes, M. Stuiver, J. W. C. White, S. J. Johnsen and J. Jouzel (1993) Comparison of the oxygen isotope records from the GISP2 and GRIP Greenland ice cores, Nature, 366: 552 - 554.

C. U. Hammer, H. B. Clausen, W. Dansgaard, A. Neftel, P. Krinstindottir and E. Johnson (1985) Continuous impurity analysis along the Dye 3 deep core, in Geophysical Monograph 33, edited by pp. 90-94,

C. U. Hammer, H. B. Clausen and C. C. J. Langwy (1994) ECM stratigraphic dating of the Byrd station ice core, Ann. Glaciol., 20: 115 - 120.

M. Hansson and E. Saltman (1993) The first Greenland ice core record of MSA and sulfate over a full glacial-interglacial cycle., Geophys.Res. letters, 23: 1163 -1166.

L. D. D. Harvey (1988) Climatic impact of ice-age aerosols., Nature, 334: 333 - 335.

S. J. Johnsen, H. B. Clausen, W. Dansgard, K. Fuhrer, N. S. Gunderstrup, C. U. Hammer, P. Iverssen, J. Jouzel, B. Stauffer and J. P. Steffensen (1992) Irregular glacial interstadials recorded in a new Greenland ice core., Nature, 359: 311-313.

S. J. Johnsen, W. Dansgaard and J. W. White (1989) The origin of Arctic precipitation under present and glacial conditions, Tellus, 41: 452-469.

S. Joussaume and J. Jouzel (1993) Paleoclimatic tracers: an investigation using an atmospheric General Circulation Model under ice age conditions. 2. Water Isotopes, J. Geophys. Res., 98: 2807-2830.

S. Joussaume, J. Jouzel and R. Sadourny (1984) A general circulation model of water isotope cycles in the atmosphere, Nature, 311: 24-29.

J. Jouzel, N. I. Barkov, J. M. Barnola, M. Bender, J. Chappelaz, C. Genthon, V. M. Kotlyakov, V. Lipenkov, C. Lorius, J. R. Petit, D. Raynaud, G. Raisbeck, C. Ritz, T. Sowers, M. Stievenard, F. Yiou and P. Yiou (1993) Extending the Vostok ice-core record of paleoclimate to the penultimate glacial period, Nature, 364: 407-412.

J. Jouzel, R. D. Koster, R. J. Suozzo and G. L. Russell (1994) Stable Water Isotope Behaviour During the LGM : A GCM analysis, J. Geophys. Res., 99: 25791-25801.

J. Jouzel and L. Merlivat (1984) Deuterium and oxygen 18 in precipitation : modeling of the isotopic effects during snow formation, Journal of Geophysical Research, 89: 11749-11757.

J. Jouzel, G. L. Russell, R. J. Suozzo, R. D. Koster, J. W. C. White and W. S. Broecker (1987) Simulations of the HDO and $H_2{}^{18}O$ atmospheric cycles using the NASA/GISS general circulation model : The seasonal cycle for present-day conditions, Journal of Geophysical Research, 92: 14739-14760.

M. Labracherie, L. D. Labeyrie, J. Duprat, E. Bard, M. Arnold, J. J. Pichon and J. C. Duplessy (1989) The last deglaciation in the Southern Ocean, Paleoceanography, 4: 629-638.

M. Legrand (1994) Ice core chemistry : Implications for our past atmosphere, in Low-Temperature chemistry of the atmosphere, edited by E. G. K. Moortgat, pp. 421 - 445, Springer-Verlag, Berlin.

M. Legrand, R. J. Delmas and R. J. Charlson (1988a) Climate forcing implications from vostok ice core sulphate data, Nature, 334: 418-420.

M. Legrand, C. Feniet-Saigne, E. S. Saltzman, C. Germain, N. I. Barkov and V. N. Petrov (1991) Ice-core record of oceanic emissions of dimethylsulphide during the last climatic cycle, Nature, 350: 144-146.

M. Legrand, C. Lorius, N. I. Barkov and V. N. Petrov (1988b) Vostok (Antarctic ice core) : atmospheric chemistry changes over the last climatic cycle (160,000 years), Atmospheric environment, 22: 317 - 331.

C. Lorius, J. Jouzel, D. Raynaud, J. Hansen and H. Le Treut (1990) Greenhouse warming, climate sensitivity and ice-core data, Nature, 347: 139-145.

C. Lorius and L. Merlivat (1977) Distribution of mean surface stable isotope values in East Antarctica. Observed changes with depth in a coastal area, in Isotopes and impurities in snow and ice. Proceedings of the Grenoble Symposium Aug./Sep. 1975, IAHS Publication edited by IAHS, pp. 125-137, IAHS, Vienna.

L. Merlivat and J. Jouzel (1979) Global climatic interpretation of the Deuterium-Oxygen 18 relationship for precipitation, J. Geophys. Res., 84: 5029-5033.

J. Oerlemans (1993) Evaluating the role of climate cooling in iceberg production and the Heinrich events, Nature, 364: 783-786.

J. J. Pichon, L. D. Labeyrie, G. Bareille, M. Labracherie, M. Duprat and J. Jouzel (1992) Surface waters temperature changes in the high latitudes of the southern ocean over the last glacial-interglacial cycle., Paleoceanography, 7: 289-318.

G. M. Raisbeck, F. Yiou, D. Bourles, C. Lorius, J. Jouzel and N. I. Barkov (1987) Evidence for two intervals of enhanced deposition in Antarctic ice during the last glacial period., Nature, 326: 273-277.

D. Rind, D. Peteet, W. Broecker, A. McIntyre and W. Ruddiman (1986) The Impact of cold North Atlantic sea surface temperatures on climate implications for the Younger Dryas cooling, Climate Dynamics :, 1: 3-33.

B. Saltzman and M. Verbitsky (1994) CO2 and glacial cycles, Nature, 367: 419.

E. S. Schwartz (1988) Are global cloud albedo and climate controlled by marine phytoplancton?, Nature, 336: 441 - 445.

T. Sowers, M. Bender, L. D. Labeyrie, J. Jouzel, D. Raynaud, D. Martinson and Y. S. Korotkevich (1993) 135 000 year Vostok - SPECMAP common temporal framework., Paleoceanography, 8: 737-766.

K. C. Taylor, G. W. Lamorey, G. A. Doyle, R. B. Alley, P. M. Grootes, P. A. Mayewski, J. W. C. White and L. K. Barlow (1993) The "flickering switch" of late Pleitocene climate change., Nature, 361: 432-436.

P. Yiou (1994) Approches de la dynamique du paléoclimat : des données et des modèles, Ph. D. thesis, Paris VI, 213 pages.

GCM STUDIES AND PARAMETERIZATION

E. Roeckner
Max-Planck-Institut für Meteorologie
Bundesstrasse 55
D-20146 Hamburg, Germany

H. Le Treut
Laboratoire de Météorologie Dynamique
Ecole Normale Superieure
24 Rue Lhomond
F-75231 Paris, Cedex 05, France

This chapter consists of three independent but related papers, 6.1, 6.2 and 6.3.

6.1 Review

E. Roeckner and H. Le Treut

Cloud parameterizations in GCMs

Cloud formation and dissipation is governed by complex interactions among dynamic, radiative, thermodynamic, microphysical, and chemical processes, most of which are inadequately understood. The spatial scales involved in cloud formation range from tenths of micrometers for cloud condensation nuclei to hundreds of kilometers for large-scale dynamics. Most of these processes cannot be resolved by the current generation of climate models with typical grid sizes of a few hundred kilometers horizontally and about one kilometer vertically. These two limitations, i.e. lack of a comprehensive theory (as opposed to radiative transfer modeling, for example) and insufficient model resolution, are prohibitive for cloud modeling from first principles. Nevertheless, stimulated by the pioneering work of Sundqvist (1978), cloud parameterizations in many of the current GCMs are built on a better physical basis than their predecessors, aiming, in particular, for a high degree of 'internal model consistency'. In fact, most of the work on cloud parameterization was, and still is, motivated by the understanding that progress can only be achieved by allowing for more and more interactions between the different model processes. Consequently, improved cloud parameterizations

NATO ASI Series, Vol. I 35
Clouds, Chemistry and Climate
Edited by Paul J. Crutzen and Veerabhadran Ramanathan

cannot be developed in isolation but must be linked to improvements of other parameterizations as well like, for example, cumulus convection or radiative transfer.

In the first attempts to include cloud effects in GCMs, zonally averaged cloudiness was prescribed externally as a function of season,latitude and height, and the optical properties were fitted to provide the correct radiation budget (e.g., Manabe et al., 1965)

Some degree of internal consistency was introduced in the so-called diagnostic cloud schemes. In diagnostic schemes, fractional cloudiness in a grid-box is calculated from empirical relationships using one or more predictors available in the GCM such as relative humidity (Smagorinsky, 1960). In the most widely used scheme of Slingo (1987), vertical velocity, static stability and convective precipitation are used in addition. Again, as in the case of prescribed clouds, the cloud optical properties are prescribed as a function of cloud type (e.g., Hansen et al., 1983; Wetherald and Manabe, 1988). Although fractional cloudiness is computed from model variables, the link with the rest of the model physics is only indirect via changes of the radiative fluxes.

The next level of sophistication was introduced by either prescribing or parameterizing the in-cloud water content (within the cloudy fraction of a grid-box) which can then be used to derive the cloud radiative properties. In the ECMWF model, for example, the cloud water content is a constant fraction of the water vapor saturation mixing ratio (Geleyn, 1981) , in the NCAR CCM2 the cloud water content is prescribed as a function of latitude and height (Kiehl et al., 1994), while in the second-generation CCC model, the cloud water content is proportional to the adiabatic liquid water content that results when saturated air is lifted vertically through a small distance (Mc Farlane et al., 1992). In the GCMs developed at the German and French Weather Services, the cloud water content is calculated from statistical models (Heise and Roeckner, 1990; Ricard and Royer, 1993). The main limitation of these diagnostic approaches is that they do not provide a link between cloud formation and the hydrological cycle.

This weakness is largely eliminated in the so-called prognostic cloud schemes which are built more or less on the work of Sundqvist (1978). In prognostic schemes, the cloud water content is calculated from the respective continuity equation taking into account large-scale and subgrid scale transport as well as sources and sinks due to condensation, evaporation and precipitation formation (Roeckner et al., 1987; Le Treut and Li,

1988; Smith, 1990). Fractional cloudiness, however, is still computed from statistical-empirical relationships although, through coupling to the water cycle, prognostic schemes are less sensitive to parameter changes than diagnostic ones. Another simplification in current prognostic schemes is that they apply only one equation for total cloud water mass (liquid and ice phase together). Since cloud microphysics and cloud optics is very different for warm and cold clouds, respectively, the phase separation is done diagnostially by using empirical relationships between cloud water content and temperature (Smith, 1990; Roeckner et al., 1991). Another lack of internal consistency in these schemes is that the cloud water transport equation does not take into account the convectively generated fraction. Also, the interaction of clouds and turbulence in the cloud-topped boundary layer deserves further attention as well as the parameterization of the cloud droplet (or ice crystal) effective radius of the respective size distribution, generally set constant with a larger value for ice clouds than for liquid clouds.

Another problem associated with prognostic cloud schemes is the calculation of advective transport. Most of the Eulerian algorithms designed for GCMs (either spectral or finite difference) are not appropriate for computing the transport of positive definite quantities with horizontal and vertical discontinuities like cloud water. Fortunately, in large-scale models, cloud water advection is generally much smaller than the condensational source term or the precipitation sink term so that cloud water advection is often neglected in GCMs as in the UKMO model for example (Smith, 1990).

The following issues have recently been addressed in the scientific community:

<u>Subgrid-scale cloudiness:</u>

In the real atmosphere, cloud formation and subsequent condensational drying takes place on scales much below the scale of a typical GCM grid-box. For example, in a case study of a midlatitude cyclone, Walcek (1994) diagnosed mid-level cloud cover of typically 20% for area-mean relative humidities between 20% and 60% at horizontal resolutions ranging between 80km and 800km. Hence, even for stratiform clouds, subgrid scale variance of relative humidity has to be considered in a realistic cloud generation algorithm. Apart from the most widely used conventional threshold technique, higher order turbulence closure schemes can be used to estimate the fluctuations of temperature and humidity and, hence, cloud fraction (e.g., Ricard and Royer, 1993). A different approach aiming primarily for internal model consistency, has been developed by Tiedtke (1993) who formulated budget equations for cloud mass and cloud volume so that both cloud water content and cloud fraction can be obtained by numerical integration of the respective prognostic equations. This scheme has been developed for and tested in the ECMWF model.

Cloud-turbulence interaction

When clouds form in the atmospheric boundary layer (ABL), the turbulence within the ABL as well as the surface fluxes of heat and humidity are primarily maintained by the transport of turbulent kinetic energy generated by cloud-top radiative cooling. Another link between clouds and turbulence is through cloud-top entrainment and generation of buoyancy by condensation. These interactions are simulated very crudely in current GCMs. A first step towards a consistent cloud-turbulence-interaction scheme was taken by Smith (1990) who introduced the condensational effect on turbulent mixing by formulating the respective fluxes in terms of cloud-conserved variables such as total water content (sum of all phases) and cloud water temperature. A similar concept is used in the Hamburg model ECHAM3 (Roeckner et al., 1992). In the latest version of this model (ECHAM4), cloud-turbulence interaction is parameterized within a higher order closure scheme (Brinkop and Roeckner, 1995). As compared to conventional schemes, the new closure provides not only a better link between the cloud and the sub-cloud layer, through transport of turbulent kinetic energy, but also between the cloud and the inversion above as a result of cloud-top entrainment.

Interaction between convective and stratiform clouds

The prognostic cloud schemes mentioned above have been designed for stratiform clouds only. Convective cloud water is either totally neglected or immediately transformed into precipitation and/or evaporated in environmental air (Tiedtke, 1989). In the most recent versions of the ECMWF and ECHAM models, however, the stratiform and convective clouds are 'unified' in the sense that the detrained fraction of the convectively generated cloud water content is used as a source term in the stratiform cloud water equation (Tiedtke, 1993). In the case of deep convection, this approach allows to simulate the formation of cumulus anvils.

Cloud microphysics

The prognostic cloud schemes currently in use are based on very simple microphysics. A logical extension of these schemes would be to add another set of prognostic equations so that the liquid and ice phases can be treated separately. A first step in this direction was taken by Lee et al. (1992) who introduced different prognostic equations for cloud liquid water and cloud ice. More recently, Fowler et al. (1994) implemented a scheme with five prognostic equations (water vapor, cloud liquid water, cloud ice, rain and snow) in the Colorado State University (CSU) GCM. Similar schemes are currently being developed for the NCAR-CCM and for the MPI-ECHAM, for example. Further refinements in these models will include the formation of cloud condensation nuclei by implementing an aerosol dynamics model and the parameterization of droplet size classes including subsequent effects on the single scattering properties.

Cloud radiative properties

In most radiative transfer schemes used in current GCMs, the most important parameters determining the cloud effect on the radiation field are cloud fraction, cloud water content (both phases) and the effective radius of the size distribution which has to be prescribed or parameterized. In most models, the effective radius is prescribed differently for water and ice clouds, respectively, or specified as a function of height varying between 5μm in the boundary layer to 40 μm at 100 hPa (Morcrette, 1991). In the Canadian Climate Centre (CCC) second-generation model (Mc Farlane et al., 1992), a parameterization is used based on empirical relationships between the effective radius and the cloud water content, differently for the liquid and ice phase, respectively. In the latest version of the ECHAM model, the effective radius for cloud doplets is formulated in terms of the predicted liquid water content, the cloud droplet number concentration and a spectral shape parameter both of which are assumed differently for maritime and continental clouds, respectively (Johnson, 1994). A refinements of this approach has been tested by using empirical relationships between the model generated sulfate aerosol mass and the cloud droplet number concentration (Boucher and Lohmann, 1995). In the long term, however, parameterizions of the cloud droplet size distribution should be derived on the basis of aerosol dynamics models.

Cloud water transport

As already mentioned above, the algorithms used for advective transport in current GCMs are not designed for spatially inhomogeneous variables like cloud water content. More appropriate schemes such as the semi-Lagrangian approach have been designed in the past und successfully applied for water vapor transport in the NCAR CCM model, for example, (Rasch and Williamson, 1991). The same approach is used for water vapor, cloud water and chemicals in the ECHAM4 model. Although more accurate than a spectral scheme, one disadvantage of the semi-Lagrangian approach is its lack of formal mass conservation.

Model validation

Most of the validation studies performed so far are based on a comparison between the simulated and observed top-of-atmosphere radiation fluxes (Kiehl and Ramanathan, 1990; Bony et al., 1992; Soden, 1992; Kiehl et al., 1994). In these studies the impact of clouds on the radiation budget is expressed by the so-called cloud radiative forcing (CRF) defined as the difference between the total flux (including all cloud effects in a certain region at a certain time) and the respective clear-sky flux (Ramanathan et al., 1989). The CRF is defined for both the shortwave (solar) component and the longwave (terrestrial infrared) component. All models are able to reproduce the observations in the sense that, for the present climate, the global mean cooling due to the negative shortwave CRF (i.e. the cloud albedo effect) is larger than the warming due to the positive longwave CRF (i.e. the cloud greenhouse effect). However, the simulated geographical patterns reveal a number of model deficiencies. In the tropics, for example, the simulated net CRF, which is close to zero according to Earth Radiation Budget Experiment (ERBE) data, all models show an often substantial net cooling (up to -30 Wm^{-2} even in the zonal mean) caused primarily by an overestimated cloud albedo effect. This problem is related to the model assumption that all cloud water is distributed homogeneously within the cloudy part of a grid-box. As shown by Davis et al. (1990), however, the albedo of an optically thick but highly nonhomogeneous cloud is systematically lower than that for a homogeneous cloud with the same cloud water content. Over the extratropical oceans, particulary during summer, the shortwave CRF and, hence, also the net CRF is systematically underestimated, probably as a result of insufficient low-level stratiform cloudiness. As shown by Gleckler et al. (1995), this deficent simulation of the geographical distribution is typical for practically all current GCMs. In some models, the implied ocean meridional heat transport, as derived from the net surface heat balance, is even equatorward in the Southern Hemisphere due to excessive cloud radiative cooling in the tropics and insufficient cooling over the southern oceans at the same time. Hence, in order to improve the oceanic forcing in coupled models, the problem of inadequate representation of subgrid-scale cloud inhomogeneities in current GCMs will have to be addressed in the future.

Although some progress has been achieved within the past decade to improve the cloud climatologies as derived from satellite data (Rossow and Schiffer, 1991) and ground-based observations (Warren et al., 1986, 1988), respectively, there are still severe uncertainties left, in particular with respect to the vertical distribution. The same applies to other cloud parameters like cloud water content (Greenwald et al., 1993) and cloud optical depth (Tselioudis et al., 1992). Hence, most of the model validation studies performed so far compare only the observed and total cloudiness. Recent model intercomparison studies (e.g., Weare and Mokhov, 1995) indicate that on a regional and seasonal scale most models are not able to adequately reproduce the observations, and this is particularly true for the marine stratocumulus regions off the west coast of America and Africa where the models systematically understimate both cloud fraction and frequency of occurrence.

As discussed in the next section (Le Treut et al.), a quantitative assessment of simulated cloudiness is hampered by the fact that the treatment of cloud inhomogeneities and vertical cloud overlap is different in the models and in the satellite data retrieval systems.

Cloud climate interactions

Cloud climate interactions are particularly relevant in the tropics. The formation of deep convective clouds with extensive optically thick cirrus anvils is observed primarily in those areas where the SST exceeds a threshold value of about 28°C (Graham and Barnett, 1987). The radiative effect of thick cirrus is a warming of the atmospheric column below cloud base through trapping of longwave radiation upwelling from the warm surface and a warming of the cloud layer by absorption of solar radiation. The surface warming by downwelling radiation emitted from cloud base is of minor importance due to the strong competition of water vapor absorption within the extremely humid boundary layer at these high temperatures. Oppositely, thick cirrus is known to reflect a substantial amount of solar radiation back to space, and this cooling relative to clear skies is felt predominantly at the surface. Ramanathan and Collins (1991) presented empirical evidence that tropical SST perturbations are accompanied by changes in the shortwave CRF that would damp the SST perturbations regionally, thus providing a natural thermostat to the system. While this type of cloud climate interaction can only be simulated with a coupled atmosphere-ocean model, the impact of the atmospheric longwave cloud radiative forcing on the tropical climate has been studied with several atmospheric GCMs (Slingo and Slingo, 1988; Randall et al., 1989, Sherwood et al., 1994, Lohmann and Roeckner, 1995; see also this chapter). These studies show that the longwave cloud radiative forcing in the tropical atmosphere is part of a positive feedback loop involving latent heat release by convection, cloud radiative heating and moisture supply by the large-scale circulation. Although smaller in magnitude than the latent heating, the cloud radiative heating is a substantial mechanisms which helps maintain the strength of the thermodynamic direct circulations such as the Hadley cell, the Walker cell, and the monsoon systems. It is also suggested (Lohmann and Roeckner, 1995) that even globally uniform SST warming may be associated with a spin-up of these circulations because the increase of the cloud optical depth in a warmer and moister atmosphere is selectively felt in convectively active areas such as the West Pacific warm pool region, for example.

Climate change and cloud-climate feedbacks

Cloud-climate feedbacks are known to be a major source of uncertainty in projections of climate change due to increased levels of greenhouse gases in the atmosphere (Mitchell et al., 1989; Cess et al., 1990). The changes of cloud parameters that may be associated with climate change (i.e., fractional cover, height, water content, droplet size distribution, phase, etc.) are individually and collectively capable of changing the Earth's radiation budget in a significant manner. The precise nature of the cloud changes and the resulting feedbacks is unknown at present because (i) only a subset of the above feedbacks is represented in current climate models, (ii) the cloud parameterizations (and most other physical parameterizations as well) are sometimes based on rather arbitrary or empirical assumptions, and (iii) a quantitative assessment of cloud feedback is difficult due to often large but partly or totally compensating effects, such as the different radiative responses in the solar and terrestrial parts of the spectrum.

The most intensively studied cloud feedback is that related to a change of cloudiness, i.e. cloud fraction and cloud height (e.g., Wetherald and Manabe, 1980; 1988; Hansen et al., 1984). Virtually all models show a decrease of global mean cloudiness in a warmer atmosphere. The models also indicate a lifting of the tropopause level with a subsequent lifting of the high-cloud layer, an increase of clouds in the polar regions and a decrease of middle and low-level clouds at low latitudes. These changes of the cloud distribution alone (keeping the cloud optical properties and other cloud parameters unchanged) result in a net positive cloud feedback which amplifies the initial warming through a positive effect on the radiation budget. A compensating effect through a possible increase of the cloud water content and, hence, increased cloud optical depth has been suggested by Somerville and Remer (1984). Climate models employing a prognostic cloud water equation (Roeckner et al., 1987, Mitchell et al., 1989, Le Treut and Li, 1991; Taylor and Ghan, 1992) show in fact a general increase of the in-cloud water content in a warmer atmosphere. However, the net feedback is a result of partly compensating processes such as the increased emissivity of high clouds (positive feedback) and the increased albedo of low clouds (negative feedback). The net effect may actually depend on the particular model assumptions as, for example, the temperature threshold used for separating the liquid phase and the ice phase (Li and Le Treut, 1992). A negative feedback through a cloud phase change from ice to liquid in a warmer atmosphere has been identified by Senior and Mitchell (1993). Since the precipitation process is assumed to be more efficient in cold clouds than in warm clouds, the phase shift results in a longer mean cloud lifetime, primarily in the middle troposphere, and this results in a negative cloud feedback. Uncertainties exist with respect to a possible change of cloud microphysics and its subsequent change of the size spectrum. For example, Platt (1989) concluded on the basis of temperature and cloud optical properties measurements in tropical cirrus clouds that current models seem to accurately predict the increase of ice water content with temperature but probably overestimate the resulting increase of infrared emissivity because they do not account for the shift of the ice crystal size spectrum to larger particles.

In conclusion, the observational and climate modeling studies on the cloud-climate feedback problem indicate that progress in this area can only be achieved if the cloud parameterizations in GCMs are built on a better physical basis which will require a combined effort through field measurements, satellite observations, cloud modeling and laboratory experiments.

6.2 Cloud vertical overlapping and cloud inhomogeneities: Their impact on model validation

H. Le Treut, G. Seze, W. Yu and M. Doutriaux

Introduction

The various data sets which are used to validate the simulation of cloud cover or cloud optical properties by global climate models are widely different in their nature. Ground based observations constitute a partly subjective appreciation of the cloud geometry. Satellite measurements represent the radiative impact of the clouds, at very different spatial scales, and for a large variety of different wavelenghts. It is often difficult to reconcile these different measurements, due to the complexity of the cloud 3-dimensional structure. This is why the validation of the Atmopheric General Circulation Models (AGCMs) remains partly qualitative (Weare and Mokhov, 1995). To proceed further and reach a truly quantitative assessment of the models, it is necessary to clarify the hypothesis relative to the cloud small-scale structure, which are made both in the models and in the cloud data retrieval systems. These hypothesis concern both the horizontal inhomogeneities of the cloud fields and the vertical overlapping of the corresponding cloud layers. We take here the example of validation studies carried out at LMD, using the ISCCP results. The ISCCP data set provides a continuous analysis of cloud cover and cloud properties, since July 83, using the visible and infrared radiometer measurements of operational meteorological satellites (Rossow and Schiffer, 1991). The comparison of the LMD GCM to ISCCP analysis was first carried out in a straightforward manner (Le Treut and Li, 1988). Remaining ambiguities, however, have lead to a more detailed investigation of the methodology used in these comparison exercices (Yu et al, 1991; Yu et al, 1995).

The role of subgrid-scale inhomogeneities: an illustration using ISCCP B2 analysis

The definition of cloudiness, or cloud cover, is far from being unambiguous. The ISCCP analyses assumes that the sky is homogeneously cloudy or clear at the size of a pixel. The statistics of those pixel properties over a larger area (2.5° x 2.5°) provide the definition of the cloud cover, as well as a diagnostics of cloud radiative properties. Within the GCM on the contrary, the subgrid-scale inhomogeneities are parameterized in a very simple statistical manner, assuming that the distribution of the water vapour within a grid-box is uniform between two values, $q-\Delta q$ and $q+\Delta q$, the part of this distribution exceeding the saturation value corresponding to the cloudy area. The information at the pixel size contained in the ISCCP B2 dataset, corresponding to Meteosat for July 83, was used by Yu et al (1991) to stress the importance of those subgrid-scale features. The infrared or visible mesurements, at the nadir, have in this case a pixel of 5 km, and they are sampled to a grid of 30 km width. Using a sampling rather than averaging procedure to reduce the information ensures that the small scale variability is preserved. The study of Yu et al (1991) shows that if we compare the scatter of the observed variances during one month within a model grid box (both in the visible and the infrared, and for a given hour of the day, in order to exclude perturbation by the diurnal cycle) to the corresponding one derived from the model results (30 results corresponding to a one-month run, without diurnal cycle) the spread in the model results is smaller than in the observations. When the observed radiances are averaged over the model grid-box, the spread of these values is then reduced and comparable to that of the model. These results suggest that in their present simplicity, the models can reproduce successfully the observed fluxes at the top of the atmosphere, but using erroneous cloud properties. This may introduce strong biases, for example, in the surface fluxes. In order to reproduce consistently the cloud cover and the cloud radiative impact, the models will require a much more elaborate treatment of the subgrid-scales.

A simple modelization of the subgrid-scale inhomogeneities from the model output and their impact on overlapping cloud layers.

The results summarized in this section (Yu et al, 1995) constitute a partial approach towards a more quantitative assessment of the model dependance on those subgrid scale features. We have tried to describe them explicitely by dividing a model grid into homogeneous cloud columns: Each of these columns occupies a specified fraction of the model grid, cloudiness may be present at different levels, but always with homogeneous properties. At each layer, the spatial average of the cloud properties gives back the prediction of the large-scale model. But dividing a grid-box within such columns depends on the cloud layer overlapping assumptions. Different cases were considered: Random overlapping, which means minimal correlation between the subgrid-scale features of two different layers, or maximum overlapping, which corresponds on the contrary to a maximal correlation. A mixed solution in which adjacent cloud layers are grouped into blocks were also considered: the clouds within a block having a maximum overlap, and the block a random overlap. A radiative calculation is done in each of these columns. We then have the equivalent of a pixel distribution, which we may process using the same ditribution classes as in the ISCCP analysis. We can therefore use the model results to obtain a more quantitative assessment of how valid the comparison to the ISCCP analysis may be. For example we show in Figure 1 the cloud distribution between different classes set up by ISCCP for the whole globe. The clouds are distinguished depending on their top temperature and optical thickness. The results shown correspond to the ISCCP data themselves, and the model results using various overlapping assumptions. We see the major impact of the cloud subgrid-scale assumption on the definition of the low cloud cover, whereas the impact on high clouds is less important. We can conclude from the comparison, however, that the model presents too much low thick clouds and too much high louds, since it exceeds the ISCCP value whatever the overlapping assumption is.

Conclusion

In summary, it is necessary to take into account the cloud subgrid structure to validate the general circulation models in a consistent manner. We have designed here a method which reproduces the ISCCP classification. The use of different cloud overlapping assumptions provides some error bar for the model data comparison exercise. Our results suggest that these diagnostics are particularly important for low clouds. They will take their full significance, however, only when the overall behaviour of the models is improved.

Figure 1a
Cloud cover distribution as a function of cloud top height from C1 data, and from model results using maximum, mix ed and random overlapping hypothesis.

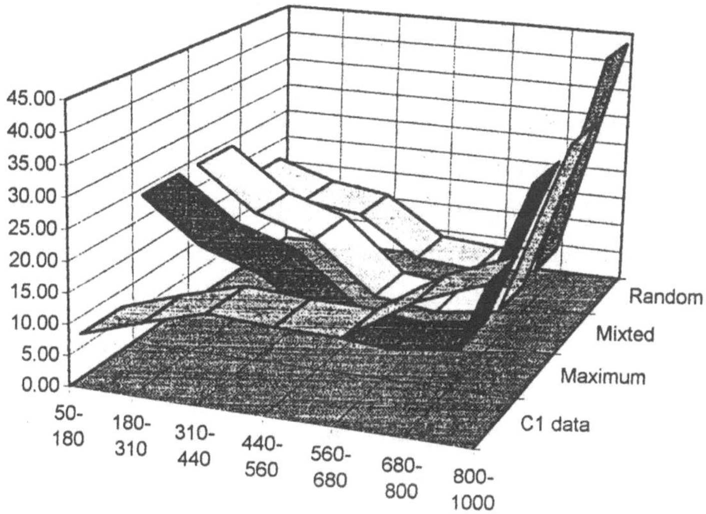

Figure 1b
Cloud cover distribution as a function of cloud optical thickness from C1 data, and from model results using maximum, mix ed and random overlapping hypothesis.

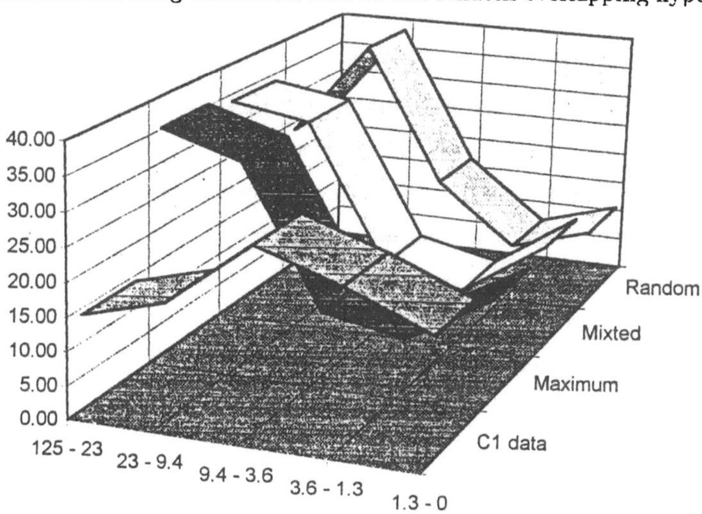

6.3 Cloud sensitivity experiments with the ECHAM model

E. Roeckner and U. Lohmann

Introduction

The general circulation of the atmosphere is ultimately driven by differential diabatic heating. On the other hand, the diabatic heat sources are strongly influenced by the circulation itself. One example of such an interplay between the diabatic heating sources and the large-scale circulation is the Walker circulation in the tropical Pacific which is forced by the SST contrast between the warm water pool in the Indonesian region and the cold water upwelling regions in the eastern part. The Walker circulation is maintained essentially by the release of latent heat in the convectively active Indonesian region which balances the adiabatic and radiative cooling in the rising branch of the cell. In the eastern Pacific, on the other hand, the radiative atmospheric cooling is balanced by adiabatic warming due to large-scale subsidence. The 'fuel' for the convective heating is provided within the lower branch of the Walker cell by moisture picked up by the trade wind systems and converging in the tropical West Pacific.

This traditional view of the Walker circulation does not take into account the radiative effect of clouds. Analyses of Earth Radiation Budget Experiment (ERBE) data indicate that thick cirrus clouds generated by deep convection may reduce the outgoing longwave radiation with respect to clear skies by up to 100 Wm^{-2} (Ramanathan and Collins, 1991), thus warming the troposphere-surface system by a substantial amount. The vertical distribution of this warming cannot be inferred from satellite data. However, modeling studies (e.g. Slingo and Slingo, 1988) suggest that in moist tropical regions most of the warming (about 80%) is felt by the atmosphere. On the other hand, the additional reflection of solar radiation by clouds is cooling primarily the Earth's surface so that in the convectively most active regions of the tropics thick cirrus anvil clouds contribute about 80 Wm^{-2} to the local warming of the troposphere. This is about 25% of the column mean latent heat release, assuming a mean precipitation rate of 10 mm/d which is typical for the Indonesian area. Thus, although certainly smaller than the latent heating, cloud radiative effects may have a non-negligible effect on the strength of the Walker circulation. In order to test this hypothesis, we have performed a series of model sensitivity experiments with different values for the cirrus cloud emissivity.

Model and experiments

A low-resolution version (T21) of the ECHAM3 model (Roeckner et al., 1992) is used with a modification of the numerical scheme: Insted of spectral advection, a semi-Lagrangian technique (Rasch and Williamson, 1991) is used for computing horizontal and vertical advection of water vapor and cloud water.The radiation scheme is based on a two-stream approximation of the radiative transfer equations with six spectral intervals in the terrestrial infrared and four in the solar part of the spectrum (Hense et al., 1982). Gaseous absorption due to water vapor, carbon dioxide and ozone is taken into account as well as scattering and absorption due to prescribed aerosol and model generated clouds. The cloud optical properties are parameterized in terms of the cloud water content (Stephens, 1978). The cloud water content is obtained from the respective budget equation including sources and sinks due to condensation, evaporation and precipitation formation by coalescence of cloud droplets and sedimentation of ice crystals (Sundqvist, 1978; Roeckner et al., 1991). The phase separation in clouds is parameterized in terms of ambient temperature on the basis of empirical data (Matveev, 1984). Sub-grid scale condensation and cloud formation is taken into account by specifying appropriate thresholds of relative humidity depending on height and convective activity (Xu and Krueger, 1991). The convection scheme comprises the effect of deep, shallow and mid-level convection on the budgets of heat, water vapor and momentum (Tiedtke, 1989). Cumulus clouds are represented by a bulk model including the effect of entrainment and detrainment on the updraft and downdraft convective mass fluxes.

Three experiments with prescribed sea surface temperatures (climatological July) have been performed over a period of 720 days each. In two experiments the infrared emissivity (ε) of all clouds above a height of 400 hPa is set to extreme values ('transparent cirrus': $\varepsilon = 0$, and 'black cirrus': $\varepsilon = 1$, respectively), while in the control experiment the cirrus emissivity is calculated as a function of the cloud water content which is a prognostic variable of the model. These three experiments have been repeated also with a globally uniform SST perturbation of 4K.

Results

The following examples demonstrate that according to this modeling study the strength of the Walker circulation is in fact crucially dependent on the magnitude of the longwave cirrus cloud radiative forcing. Although the longwave radiative heating through cirrus clouds in the troposphere is nearly an order of magnitude smaller than the latent heat release, it has a significant impact on convective activity and large-scale dynamics. If its relatively small direct contribution to the total diabatic heating of about 10% is removed, as in the transparent-cirrus experiment, the total diabatic heating in the convectively active West Pacific warm pool area and over the Indian subcontinent is reduced by a much larger amount of typically 50%. In the nearly cloud-free East Pacific, the diabatic cooling is significantly reduced as well, and the divergent part of the Walker circulation is about 50% weaker than in the control experiment. On the other hand, the Walker circulation is increased by roughly 50% relative to the control experiment if cirrus clouds are treated as black bodies, and a reverse Walker cell forms in the upper troposphere and lower stratosphere. A similar but weaker response is simulated in the global warming experiment. The fuel for the enhanced convection in the disturbed regions is provided by additional moisture convergence through the enhanced Hadley and Walker circulations while the evaporation over the tropical oceans actually decreases with increasing cirrus emissivity as a result of the warmer and moister boundary layer.

Figures 1 to 3 show long term mean (perpetual July, day 91-720) longitude-height cross sections (average 0°-30°N) of zonal wind , velocity potential and total diabatic heating, respectively. In the top panel the transparent cirrus (TC) experiment is shown, in the middle the control (variable cirrus, VC) and in the bottom the black cirrus case (BC). There is a substantial increase of the zonal wind (Fig. 1) when the cirrus emissivity is increased from zero (top) to one (bottom), and the patterns of convergence and divergence change in the same sense (Fig. 2). The response of the total diabatic heating (Fig. 3) is much larger than the initial radiative perturbation which is generally less than 0.5 K/d, even within the convectively most active region between 60°E and the dateline. Most of the differences result actually from the weaker (larger) moisture supply due to the weaker (stronger) Walker circulation and less (more) active convection in the TC (BC) experiment. Figure 4 demonstrates that even the global precipitation patterns

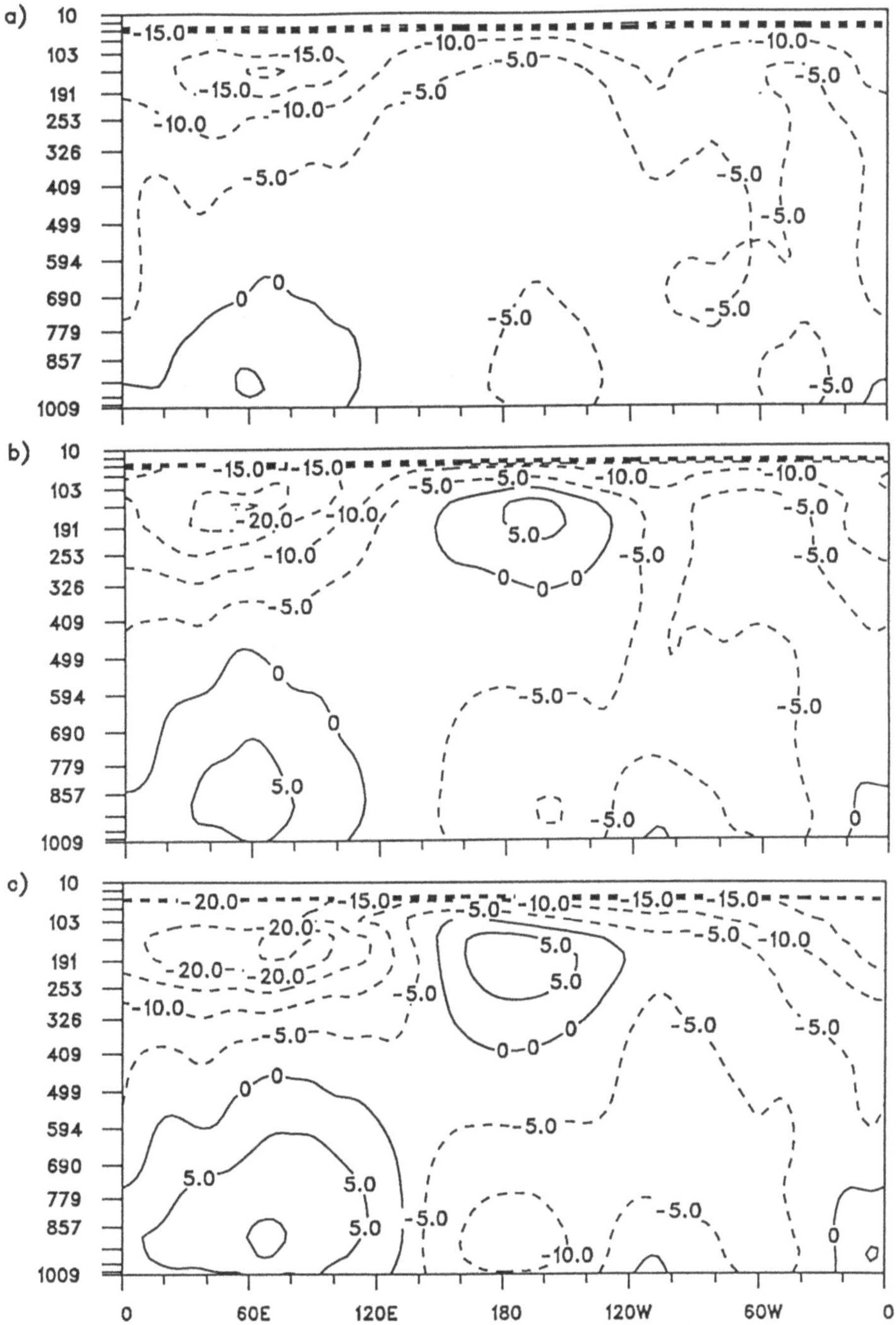

Figure 1. Longitude-pressure (hPa) cross section of zonal wind (m/s) averaged between 0° and 30° N for day 91-720 of a perpetual July experiment. a) transparent cirrus, b) variable cirrus emissivity (function of predicted cloud water path), and c) black cirrus

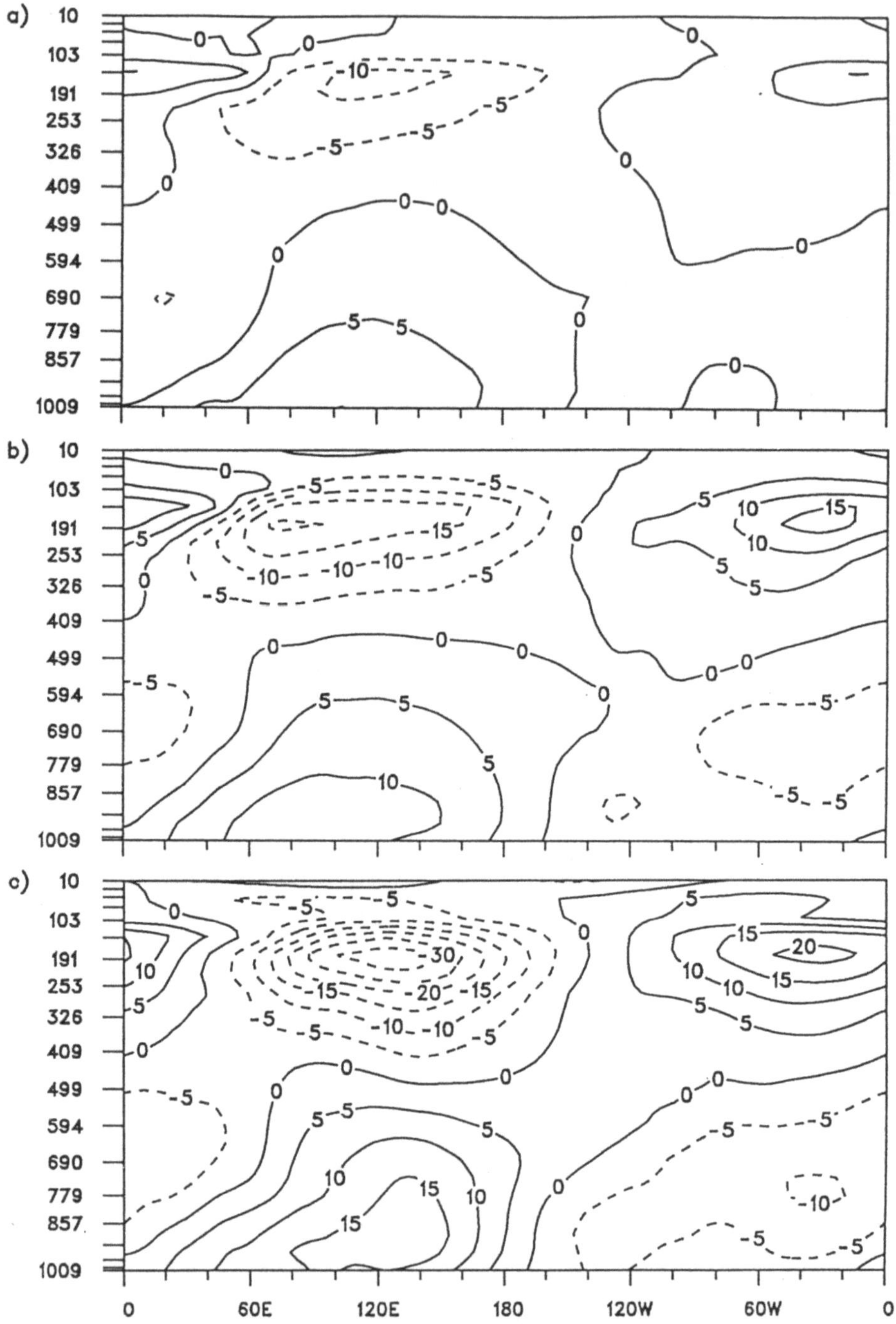

Figure 2. Longitude-pressure (hPa) cross section of velocity potential (km^2/s) averaged between 0° and 30° N for day 91-720 of a perpetual July experiment. a) transparent cirrus, b) variable cirrus emissivity (function of predicted cloud water path), and c) black cirrus

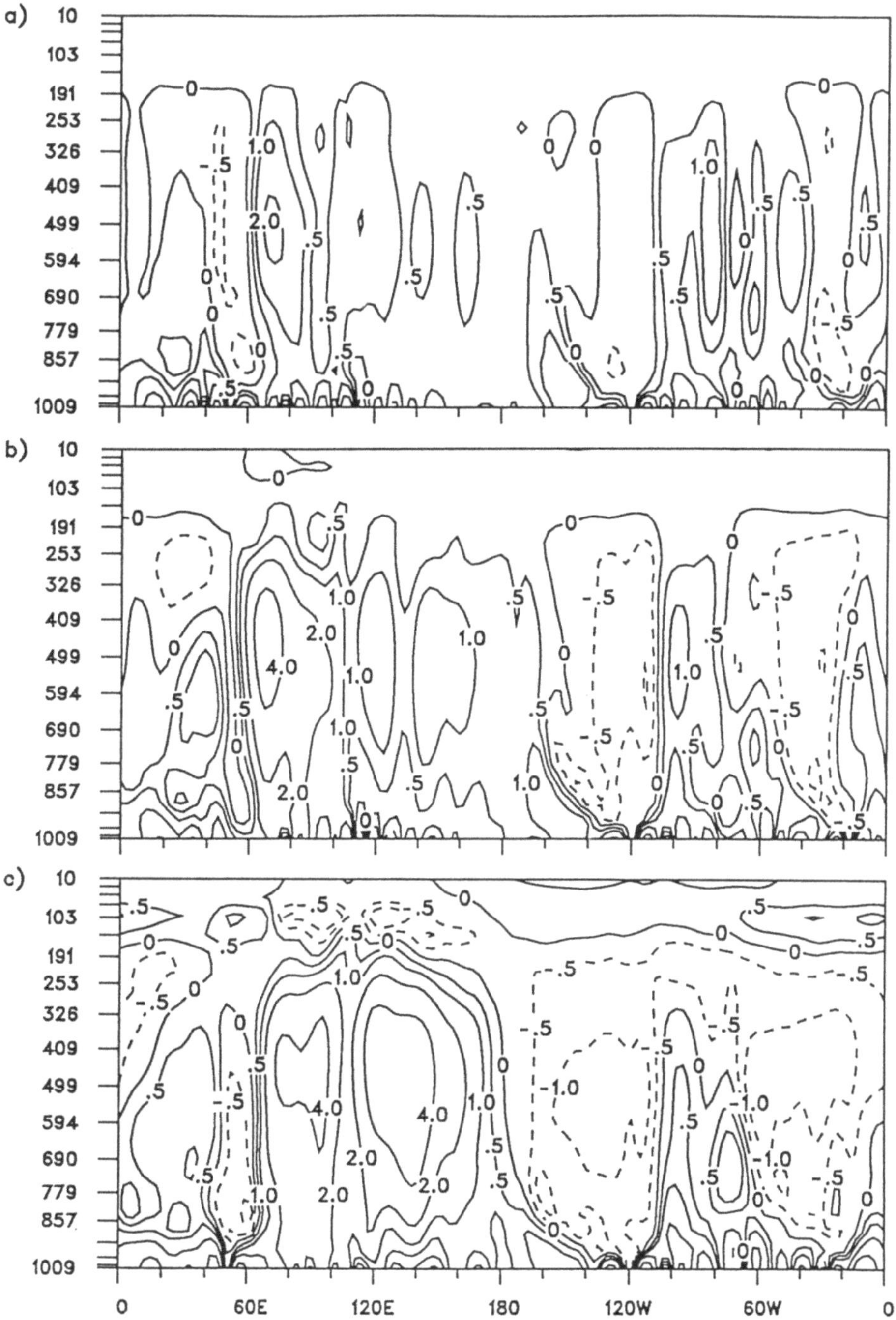

Figure 3. Longitude-pressure (hPa) cross section of total diabatic heating (K/d) averaged between 0° and 30° N for day 91-720 of a perpetual July experiment. a) transparent cirrus, b) variable cirrus emissivity (function of predicted cloud water path), and c) black cirrus

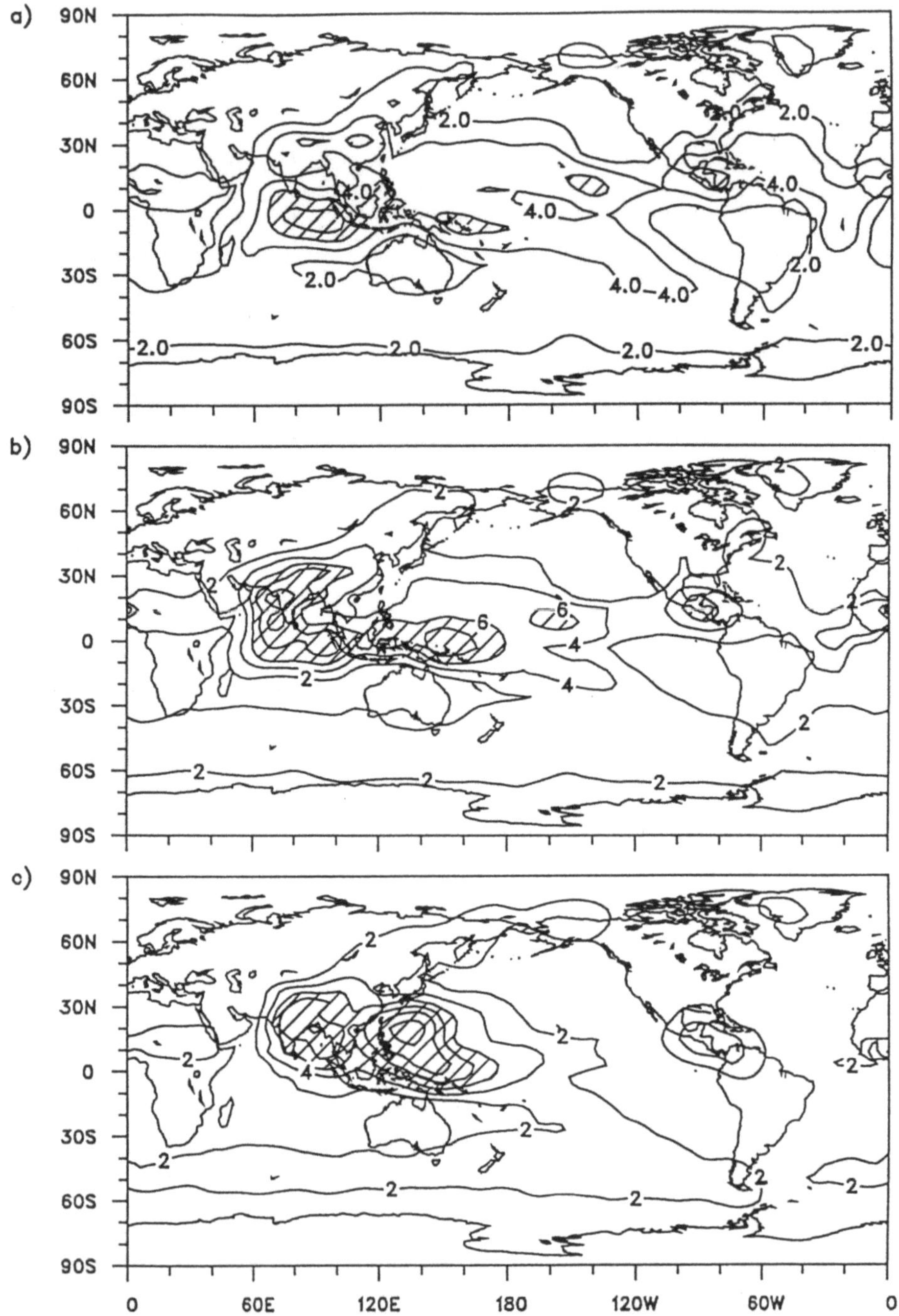

Figure 4. Global distribution of precipitation (mm/d) for day 91-720 of a perpetual July experiment. a) transparent cirrus, b) variable cirrus emissivity (function of predicted cloud water path), and c) black cirrus

change significantly when the cirrus emissivity is set to the extremes. In the TC run (Fig. 4a) the precipitation is more widespread while the maximum values are smaller than in the control. The opposite holds for the BC experiment (Fig. 4c) where the peak values are larger while the precipitation is reduced in most other parts of the world and also in the global mean.

Conclusion

This study indicates that the tropical circulation is not only driven by latent heating, although this is by far the largest term in the diabatic heat budget. The contribution from cirrus cloud radiative heating is actually much larger than a comparison with the convective heating rate in the control experiment would suggest. These results support the main conclusion from previous GCM experiments (Slingo and Slingo, 1988; Randall et al., 1989) that cirrus clouds play a key role in maintaining the tropical circulation through a positive feedback loop between cloud radiative heating, deep cumulus convection and large-scale moisture supply. Since these three models use different parameterizations for radiation, cloud formation, cumulus transport and turbulent transfer, it seems fair to conclude that the respective closure assumptions do not play a decisive role in generating these interactions.

According to the SST perturbation experiments, there is evidence, furthermore, that the radiative-convective-dynamical feedback in the tropics might be important for the global warming problem where a larger atmospheric moisture content through enhanced surface evaporation will provide additional fuel to drive the Hadley and Walker circulations, even in the case of spatially uniform SST warming.

References

Bony, S., H. Le Treut, J.-P. Duvel and R.S. Kandel, 1992: Satellite validation of GCM-simulated annual cycle of the earth radiation budget and cloud forcing. J.Geophys.Res., 97, D16,18061-18081.

Boucher, O. and U. Lohmann, 1995, The sulfate-CCN-cloud albedo effect: A sensitivity study with two general circulation models. Tellus, in press.

Brinkop, S., and E. Roeckner, 1995: Sensitivity of a general circulation model to parameterizations of cloud-turbulence interactions in the atmospheric boundary layer. Tellus, in press.

Cess, R.D., G.L. Potter, J.P. Blanchet, G.J. Boer, A.D. Del Genio, M. Déqué, V. Dymnikov, V. Galin, W.L. Gates, S.J. Ghan, J.T. Kiehl, A.A. Lacis, H. Le Treut, Z.-X. Li, X.-Z. Liang, B.J. McAvaney, V.P. Meleshko, J.F.B. Mitchell, J.-J. Morcrette, D.A. Randall, L. Rikus, E. Roeckner, J.F. Royer, U. Schlese, D. A. Sheinin, A. Slingo, A.P. Sokolow, K.E. Taylor, W.M. Washington, R.T. Wetherald, I. Yanai and M.-H. Zhang, 1990: Intercomparison and interpretation of climate feedback processes in 19 atmospheric general circulation models. J.Geophys.Res., 95, 16601-16615.

Davis, A., P. Gabriel, S. Lovejoy, D. Schertzer and G.L. Austin, 1990: Discrete angle radiative transfer: 3. Numerical results and meteorological applications. J.Geophys.Res., 95, 11729-11742.

Fowler, L.D., D.A. Randall and S.A. Rutledge, 1994: Liquid and ice cloud microphysics in the CSU general circulation model. Part I: Model description and simulated microphysical processes. Submitted to J. Climate.

Geleyn, J.-F., 1981: Some diagnostics of the cloud/radiation interaction in the ECMWF forecasting model. ECMWF Workshop on Radiation and Cloud-Radiation Interaction in Numerical Modelling, 135-162, 15-17 October 1980, ECMWF, Reading, UK.

Gleckler, P.J., D.A. Randall, G. Boer, R. Colman, M. Dix, V. Galin, M. Helfand, J. Kiehl, A. Kitoh, W. Lau, X.-Z. Liang, V. Lykossov, B. McAvaney, K. Miyakoda, S. Planton and W. Stern, 1995: Interpretation of ocean energy transports implied by atmospheric general circulation models. Submitted to Geophys.Res.Lett.

Graham, N.E. and T.P. Barnett, 1987: Sea surface temperature, surface wind divergence, and convection over tropical oceans. Science, 238, 657-659.

Greenwald, T.J., G.L. Stephens, T.H. Vonder Haar and D.L. Jackson, 1993: A physical retrieval of cloud liquid water over the global oceans using Special Sensor Microwave/Imager (SSM/I) observations. J.Geophy.Res., 98, 18471-18488.

Hansen, J., G. Russell, D. Rind, P. Stone, A. Lacis, S. Lebedeff, R. Ruedy and L. Travis, 1983: Efficient three-dimensional global models for climate studies: Models I and II. Mon.Wea.Rev., 111, 609-662.

Hansen, J., A. Lacis, D. Rind, G. Russell, P. Stone, I. Fung, R. Ruedy and J. Lerner, 1984: Climate sensitivity: Analysis of feedback mechanisms. Climate Processes and Climate Sensitivity. J. Hansen and T. Takahashi, Eds., Geophys.Monogr., No. 29, Amer.Geophys.Union, 130-163.

Heise, E. and E. Roeckner, 1990: The performance of physically based cloud schemes in general circulation models. Beitr.Phys.Atmos., 63, 1-14.

Hense, A., M. Kerschgens and E. Raschke, 1982: An economical method for computing radiative transfer in circulation models. Quart. J. Roy. Meteor. Soc., 108, 231-252.

Johnson, D.W., 1993: Parameterisation of the cloud topped boundary layer: Aircraft Measurements. ECMWF Workshop Proceedings 'Parameterization of the cloud topped boundary layer', 77-117, European Centre for Medium-Range Weather Forecasts, Reading, UK.

Kiehl, J.T. and V. Ramanathan, 1990: Comparison of cloud forcing derived from the Earth Radiation Budget Experiment with that simulated by the NCAR Community Climate Model. J.Geophys.Res., 95, 11679-11698.

Kiehl, J.T., J.J. Hack and B.P. Briegleb, 1994: The simulated Earth radiation budget of the National Center for Atmospheric Research community climate model CCM2 and comparisons with the Earth Radiation Budget Experiment (ERBE). J.Geophys. Res., 99, D10, 20815-20827.

Lee, J.L., K.N. Liou and S.C. Ou, 1992: A three-dimensional large-scale cloud model: Testing the role of radiative heating and ice phase processes. Tellus, 44A, 197-216.

Le Treut, H. and Z.X. Li, 1988: Using Meteosat data to validate a prognostic cloud generation scheme. Atmos.Res., 21, 273-292.

Le Treut, H. and Z.-X. Li, 1991: Sensitivity of an atmospheric general circulation model to prescribed SST changes: Feedback effects associated with the simulation of cloud optical properties. Climate Dyn., 5, 175-187.

Li, Z.-X. and H. Le Treut, 1992: Cloud-radiation feedbacks in a general circulation model and their dependence on cloud modelling assumptions. Climate Dyn., 7,133-139.

Lohmann, U. and E. Roeckner, 1995: The influence of cirrus cloud-radiative forcing on climate and climate sensitivity in a general circulation model. Submitted to J.Geophys.Res.

Manabe, S., J. Smagorinsky and R.F. Strickler, 1965: Simulated climatology of a general circulation model with a hydrological cycle. Mon.Wea.Rev., 93, 769-798.

Matveev, L.T., 1984: Cloud dynamics. Atmospheric Sciences Library. D. Reidel Publishing Company, Dordrecht, 340 pp.

McFarlane, N.A., G.J. Boer, J.-P. Blanchet and M. Lazare, 1992: The Canadian Climate Centre second-generation general circulation model and its equilibrium climate. J.Climate, 5, 1013-1044.

Mitchell, J.F.B., C.A. Senior and W.J. Ingram, 1989: CO_2 and climate: A missing feedback? Nature, 341, 132-134.

Morcrette, J.-J., 1991: Radiation and cloud radiative properties in the European Centre for Medium Range Weather Forecasts forecasting system. J.Geophys.Res., 96, 9121-9132.

Platt, C.M.R., 1989: The role of cloud microphysics in high-cloud feedback effects on climate change. Nature, 341, 428-429.

Ramanathan, V., R.D. Cess, E.F. Harrison, P. Minnis, B.R. Barkstrom, E. Ahmad and D. Hartmann, 1989: Cloud-radiative forcing and climate: Results from the Earth Radiation Budget Experiment. Science, 243, 57-63.

Ramanathan, V. and W. Collins, 1991: Thermodynamic regulation of ocean warming by cirrus clouds deduced from the 1987 El Nino. Nature, 351, 27-32.

Randall, D.A., Harshvardhan, D.A. Dazlich and T.G. Corsetti, 1989: Interactions among radiation, convection and large-scale dynamics in a general circulation model. J.Atmos.Sci., 46, 1943-1970.

Rasch, P.J. and D.L. Williamson, 1991: The sensitivity of a general circulation model climate to the moisture transport formulation. J.Geophys.Res., 96, D7, 13123-13137.

Ricard, J.L. and J.F. Royer, 1993: A statistical cloud scheme for use in an AGCM. Ann.Geophysicae, 11, 1095-1115.

Roeckner, E., U. Schlese, J. Biercamp and P. Löwe, 1987: Cloud optical depth feedbacks and climate modelling. Nature, 329, 138-140.

Roeckner, E., M. Rieland and E. Keup, 1991: Modelling of cloud and radiation in the ECHAM model. ECMWF/WCRP Workshop on "clouds, radiative transfer and the hydrological cycle", 12-15 Nov. 1990, 199-222, ECMWF, Reading, U.K.

Roeckner, E., K. Arpe, L. Bengtsson, S. Brinkop, L. Dümenil, M. Esch, E. Kirk, F. Lunkeit, M. Ponater, B. Rockel, R. Sausen, U. Schlese, S. Schubert and M. Windelband, 1992. Simulation of the present-day climate with the ECHAM model: Impact of model physics and resolution. Max-Planck-Institut für Meteorologie, Report No. 93, 171 pp: Hamburg.

Rossow, W. and R.A Schiffer, 1991: ISCCP cloud data products. Bull.Am.Meteorol.Soc., 72, 2-20.

Senior, C.A. and J.F.B. Mitchell, 1993: Carbon dioxide and climate: The impact of cloud parameterization. J.Climate, 6, 393-418.

Sherwood, S.C., V. Ramanathan, T.P. Barnett, M.K. Tyree and E. Roeckner, 1994: Response of an atmospheric general circulation model to radiative forcing of tropical clouds. J.Geophys.Res., 99, D10, 20829-20845.

Slingo, J.M., 1987: The development and validation of a cloud cover prediction scheme for the ECMWF model. Q.J.R.Meteorol.Soc., 113, 899-927.

Slingo, A. and J.M. Slingo, 1988: The response of a general circulation model to cloud longwave radiative forcing. I. Introduction and initial experiments. Q.J.R.Meteorol. Soc., 114, 1027-1062.

Smagorinsky, J., 1960: On the dynamical prediction of large-scale condensation by numerical methods. Geophys.Monogr., 5, 71.

Smith, R.N.B. 1990. A scheme for predicting layer clouds and their water content in a general circulation model. Q.J.R.Meteorol.Soc. 116, 435-460.

Soden, B.J., 1992: Validation of cloud forcing simulated by the National Center for Atmospheric Research Community Climate Model using observations from the Earth Radiation Budget Experiment. J.Geophys.Res., 97, D16, 18137-18159.

Somerville, R.C.J. and L.A. Remer, 1984: Cloud optical thickness feedbacks in the CO_2 climate problem. J.Geophys.Res., 89, 9668-9672.

Stephens, G.L., 1978: Radiation profiles in extended water clouds, Part 2: Parameterization schemes. J. Atmos. Sci., 35, 2123-2132.

Sundqvist, H. 1978. A parameterization scheme for non-convective condensation including prediction of cloud water content. Q.J.R.Meteorol.Soc., 104, 677-690.

Taylor, K.E. and S.J. Ghan, 1992: An analysis of cloud liquid water feedback and global climate sensitivity in a general circulation model. J.Climate, 5, 907-919.

Tiedtke, M. 1989. A comprehensive mass flux scheme for cumulus parameterization in large-scale models. Mon.Wea.Rev., 117, 1779-1800.

Tiedtke, M., 1993: Representation of clouds in large-scale models. Mon.Wea.Rev., 121, 3040-3061.

Tselioudis, G., W. Rossow and D. Rind, 1992: Global patterns of cloud optical thickness variation with temperature. J.Climate, 5, 1484-1495.

Walcek, C.J., 1994: Cloud cover and its relationship to relative humidity during a springtime midlatitude cyclone. Mon.Wea.Rev., 122, 1021-1035.

Warren, S. G., Hahn, C.J., London, J., Chervin, R.M. and Jenne, R.L. 1986. Global distribution of total cloud cover and cloud type amounts over land. NCAR Tech. Note, NCAR TN-273 + STR, Boulder Colo., and DOE Tech. Rep. ER/60085-H1, 29pp., 199 maps, U.S. Dept. of Energy, Washington, D.C., U.S.A.

Warren, S.G., Hahn, C.J., London, J., Chervin, R.M. and Jenne, R.L. 1988. Global distribution of total cloud cover and cloud type amounts over the ocean. NCAR Tech. Note, NCAR TN-317+ STR, Boulder Colo., and DOE Tech. Rep. ER/0406, 42pp., 170 maps, U.S. Dept. of Energy, Washington, D.C., U.S.A.

Weare B.C. and I.I. Mokhov, and AMIP Modelling groups, 1995: Evaluation of total cloudiness and its variability in the Atmospheric Model Intercomparison Project. Submitted to J. Climate.

Wetherald, R.T. and S. Manabe, 1980: Cloud cover and climate sensitivity. J. Atmos. Sci., 37, 1485-1510.

Wetherald, R.T. and S. Manabe, 1988: Cloud feedback processes in a general circulation model. J.Atmos.Sci., 45, 1397-1415.

Xu, K.M. and S.K. Kruger, 1991: Evaluation of cloudiness parameterizations using a cumulus ensemble model. Mon. Wea. Rev., 119, 342-367.

Yu, W., G. Seze, H. Le Treut and M. Desbois, 1991: Comparison of radiance fields observed by satellite and simulated by the LMD GCM, Dyn.Atm.Ocean, 16, 147-165.

Yu, W., M. Doutriaux, G. Sèze, H. Le Treut and M. Desbois, 1995: A methodological study of the validation of cloud water schemes in a GCM (in preparation).

THE CENTRAL EQUATORIAL PACIFIC EXPERIMENT

W.D. Collins[1], V. Ramanathan[1], P.J. Crutzen[1,2], A. Heymsfield[1,3], J. P. Kuettner[3], D. Kley[1,4], R. L. Grossman[5]

[1]Center for Clouds, Chemistry and Climate
Scripps Institution of Oceanography, La Jolla, CA 92093-0239, USA

[2]Max-Planck-Institut für Chemie
D-55020 Mainz, Germany

[3]National Center for Atmospheric Research
Boulder, CO 80307, USA

[4]Department für Chemie und Dynamik der Geosphäre
Forschungszentrum Jülich, 52425 Jülich, Germany

[5]Astrophysical, Planetary and Atmospheric Sciences
University of Colorado, Boulder, CO 80309, USA

1 Overview and scientific framework for CEPEX

1.1 Introduction

The tropical Pacific Ocean exhibits several intriguing features. For nearly 50% of the tropical Pacific (30°N to 30°S), sea-surface temperatures (SSTs) are within a narrow range between 300 K and 303 K, and less than 0.01% of the ocean surface has SSTs in excess of 304 K (Figure 1). Furthermore, deep convection is triggered only when SSTs exceed 300 K. In regions of deep convection and warm SST, the reduction of outgoing longwave energy by the atmospheric water vapor increases so rapidly with SST that it exceeds the rate of increase of blackbody surface emission. Thus the surface-atmosphere system in the warm pool loses its ability to radiate excess energy to space. Without some negative feedback process, the greenhouse effect would produce a runaway warming. *Ramanathan and Collins* [1991] (hereafter RC) found that the warm ocean is also covered by thick cirrus anvil clouds, which reflect a significant amount of solar energy back to space. They concluded that the anvils produced by deep convection act like a thermostat to regulate the flow of solar energy to the ocean.

In order to test the thermostat hypothesis, the fluxes and state parameters of the ocean-atmosphere system that enter into the hypothesis should be measured as a function of SST over the equatorial ocean. In the equatorial Pacific, the transition between the warm pool and the colder eastern ocean is located close to the international dateline. To the west, the higher temperatures are accompanied with deep convection, which can be readily detected by satellite. To the east, the temperatures are lower because of upwelling driven by the prevailing easterlies. Deep convection in the eastern equatorial

NATO ASI Series, Vol. I 35
Clouds, Chemistry and Climate
Edited by Paul J. Crutzen and Veerabhadran Ramanathan

region is much less frequent. The east-west gradient in SST provides a natural framework to sample the transition from a cold, convectively suppressed to a warm, convectively active regime. The primary experimental objectives of CEPEX have been formulated in terms of this gradient:

- Measure the east-west gradients of the vertical distribution of water vapor along the equatorial Pacific.
- Measure, by direct atmospheric observations, the vertical structure of the water vapor greenhouse effect.
- Measure the effect of widespread cirrus layers on radiative fluxes over the equatorial Pacific.
- Measure the east-west gradients of SST and the evaporative and sensible heat fluxes from the sea surface along the equatorial Pacific.
- Explore the microphysical factors contributing to the high albedo of widespread cirrus layers.

The connection between these measurements and SST regulation is first explored within the context of a simple model. Some of the initial results related to the CEPEX objectives are presented in the remainder of the chapter.

1.2 A model for the thermostat

The thermostat hypothesis may be formulated in terms of the time-dependent equation for the temperature T_s of the ocean mixed layer in the warm pool. In its simplest form, the equation for the evolution of T_s is

$$\rho C_p D \frac{dT_s}{dt} = H + F_o \tag{1}$$

Here ρ and C_p are the density and specific heat of water, respectively, and D is the depth of the mixed layer (which will be held constant in the derivations that follow). The term H is the net flux at the ocean surface, and F_o is the total dynamical heat transport by advection, entrainment, and upwelling into or out of the mixed layer. The net surface flux may be written in terms of its thermodynamic and radiative components as

$$H = \left[\overbrace{(S_a^* + C_s^*)}^{SW} + \overbrace{\left(G_a^* + C_l^* - \sigma T_s^4\right)}^{LW} + \overbrace{(F_S + F_L + F_P)}^{NR} \right] \tag{2}$$

The individual terms are:

SW: Shortwave fluxes:

$$\begin{aligned} S_a^* &= \text{Clear-sky net solar flux} \\ C_s^* &= \text{Solar cloud forcing} \end{aligned}$$

LW: Longwave fluxes:

$$\begin{aligned} G_a^* &= \text{Clear-sky greenhouse effect} \\ C_l^* &= \text{Longwave cloud forcing} \\ \sigma T_s^4 &= \text{Thermal emission from sea surface} \end{aligned}$$

NR: Non-radiative fluxes

$$\begin{aligned} F_S &= \text{Sensible heat flux} \\ F_L &= \text{Latent heat flux} \\ F_P &= \text{Precipitation heat flux} \end{aligned}$$

The terms that add energy to the mixed layer are positive, i.e., S_a^*, G_a^*, and C_l^*. Terms that reduce the amount of energy, or cool, the layer are negative, i.e., C_s^*, σT_s^4, and all the non-radiative terms.

Over tropical oceans, the frequency of deep convection is related to SST, although the physical mechanisms for this relation are still being debated. The frequency of convection is relatively small when T_s is less than approximately 300 K, and increases rapidly with T_s for temperatures between 300 K and 303 K [*Graham and Barnett,* 1987; *Waliser and Graham,* 1993; *Zhang,* 1993]. Evidence from CEPEX and other field experiments suggest that most of the variability in the radiative terms of H is associated with deep convection. In the warm pool, for example, *Young et al.* [1992] find that H varies from greater than 100 W m^{-2} under clear-sky conditions to less than -30 W m^{-2} under convective conditions. Because the frequency of convection and its effects on the surface fluxes increase with SST, we expand H as a function of SST:

$$H(T_s) = H_0 + \frac{dH}{dT_s}(T_s - T_0) + \frac{1}{2}\frac{d^2 H}{dT_s}(T_s - T_0)^2 + \dots \tag{3}$$

T_0 is a threshold temperature, approximately 300 K, below which deep convection and the associated cirrus anvil clouds are infrequent. H_0 is the heat budget (eq. [2]) evaluated at $T_s \simeq T_0$. The decrease in the surface heat budget associated with the increase in convection with SST implies that $dH/dT_s < 0$. Inserting this expansion of H into the evolution equation (1) and neglecting nonlinear terms in T_s gives

$$\alpha \frac{dT_s}{dt} = H_0 + F_o - \beta(T_s - T_0) \tag{4}$$

where

$$\alpha = \rho C_p D$$
$$\beta = -\left\{\frac{dH}{dT_s}\right\} \tag{5}$$

The solution for the SST is

$$T_s = T_0 + \frac{(H_0 + F_o)}{\beta}\left[1 - \exp\left(-\frac{t}{\tau_e}\right)\right] \tag{6}$$

The e-folding time-scale is given by $\tau_e = \alpha/\beta$. The asymptotic solution in the limit when $t \gg \tau_e$, $T_s = T_0 + (H_0 + F_o)/\beta$, is the maximum equilibrium SST. The feedback term β determines both the magnitude of the maximum SST and the time-scale for changes in the mixed layer.

The evolution of the temperature from eq. (6) is shown in Figure 2 for several values of β. The parameters entering into the solutions are given in Table 1. The magnitude of the ocean flux F_o is less than 20 W m^{-2} [*Niiler and Stevenson,* 1982] and has been omitted from the calculation of T_s. The largest term in β is associated with the cloud shading effect. The maximum SST when all the feedback terms are included is approximately 304 K. When the evaporative cooling or shortwave cloud forcing feedbacks are omitted, the maximum SST is larger than that observed in the warm pool. Note that the solution without cloud shading, which corresponds to the evaporative mechanism proposed by *Newell* [1979] and others, rapidly diverges from observed SSTs.

	H_0 (W m^{-2})	D (m)	β (W m^{-2}K^{-1})	T_{max} (K)	τ_e (Days)
a	+100	30	-25	304	54
b	+100	30	-15	307	90
c	+100	30	-2.5	340	540

Table 1: Parameters for the solution of T_s. The ocean flux F_o has been set to 0 W m^{-2}, and the shortwave, longwave, and nonradiative components of β corresponding to the three components of H are +22.5, −7.5, and +10.0 W m^{-2} K^{-1}, respectively.

The e-folding time for the mixed layer when all the feedbacks are included is about 54 days. The large variations in convection associated with the Madden-Julian Oscillation have roughly the same periodicity. The MJO "interrupts" the smooth evolution

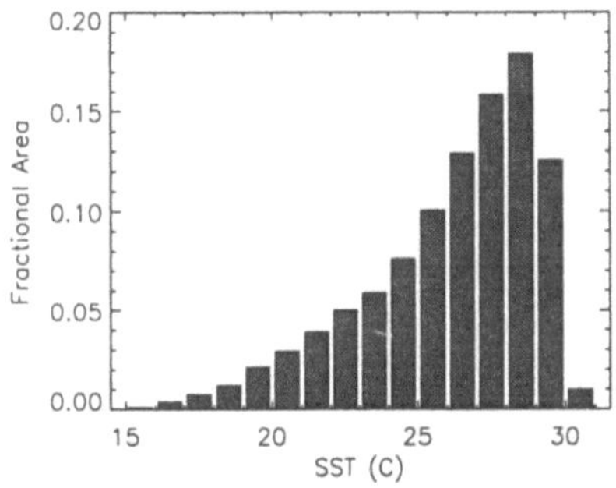

Figure 1: Spatial frequency distribution of SSTs in the tropical Pacific for 1985. The SST data are monthly mean observed values for 2° latitude by 2° longitude bins. All 12 months of the SST data from 1985 are used.

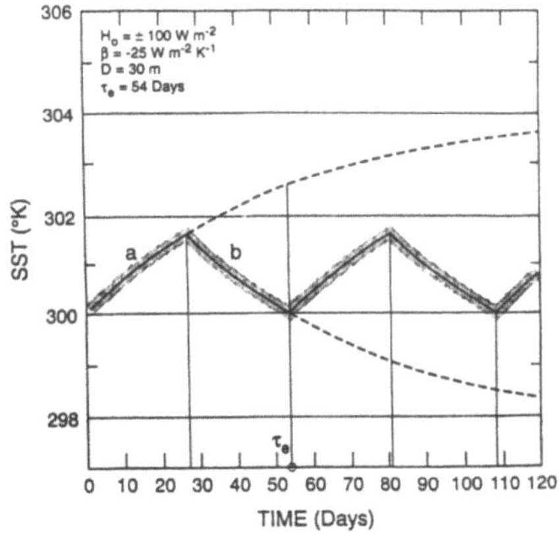

Figure 3: Evolution of T_s as a function of time including the variation of H with changing weather conditions. The parameters are the same as solution (a) (Table 1) except that the sign of H_0 is changed every $\tau_e/2$ days.

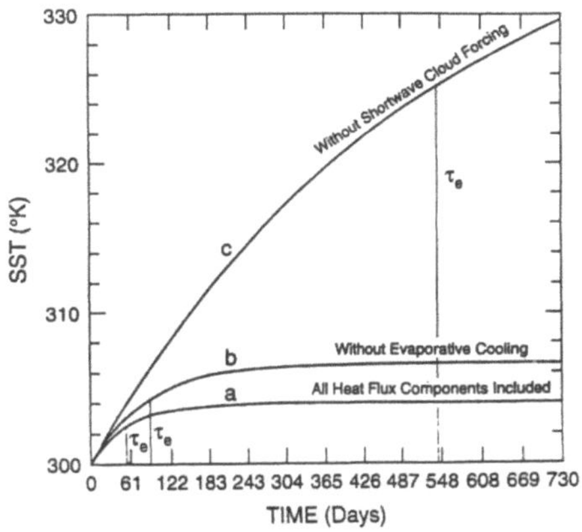

Figure 2: Evolution of T_s as a function of time from eq. (6) for three different values of β. Parameters for the solutions are given in Table 1.

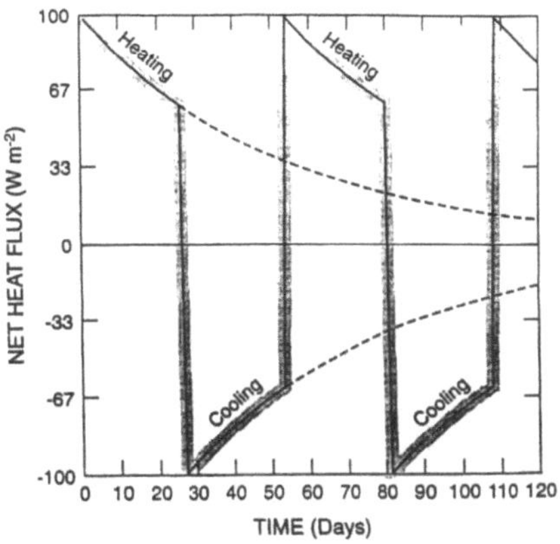

Figure 4: Net heat flux H from eq. (3) and Fig. 3.

of the SST given by eq. (6), since H changes abruptly at the onset and termination of the convective phase. The qualitative effect of the MJO is to change the sign of H_0 in the transition from clear to cloudy conditions. In order to explore the potential effects of weather events such as MJOs on the warm pool SSTs, we model the heat flux term, H_0, as a step function term, depending on the SST. When SST increases with time and reaches a pre-specified value (about 302 K in the example shown here), we change H_0 to -100 W m^{-2} to mimic a convective event. When the SST begins to cool in response to this cooling event, we set the heating back to 100 W m^{-2} when the SST dips below 300 K. The evolution of the SST including the effects of the MJO and other weather events is shown in Figure 3. The corresponding net heat flux into the mixed layer is shown in Figure 4. These results show, in a qualitative sense, the interaction between SST, weather, and the onset of convection at 300 K.

A complete analysis to determine the relation of the surface heat budget, convection, and SST is underway. Some of the preliminary findings from CEPEX are presented in the next three sections.

2 Water vapor and ozone observations

2.1 Introduction

Water vapor is the single most important radiatively active species in the troposphere. It is responsible for most of the greenhouse effect, followed by carbon dioxide and gases such as methane, nitrous oxide, ozone and others. The super greenhouse effect observed by RC and the corresponding enhancement of greenhouse heating of the ocean surface [*Vonder Haar,* 1986] are caused by water vapor. Although numerous observations suggest that convection has a significant effect on the distribution of water vapor, to date it has not been shown from direct measurements whether deep convection moistens or dries the atmosphere.

The uncertainty regarding the effects of convection is reflected in theoretical attempts to explain the observed moisture distribution. *Sun and Lindzen* [1993], for example, have modeled the vertical structure of the tropical tropospheric relative humidity over the domain of the Hadley circulation. In their model, the humidity reaches its minimum value of approximately 15% at 8 km. Above 8 km, the humidity increases to a local maximum of about 30% at 14 km. The maximum is caused by evaporation of hydrometeors carried aloft by the rising turrets of deep convection. Sun and Lindzen compare their model results to zonal means of relative humidity observed by satellite and the conventional radiosonde network. They conclude that the observations and the model are in reasonable agreement. However, Sun and Lindzen note that satellite instruments cannot measure below the cloud top, and furthermore that the profiles of humidity derived from radiosonde data in the upper troposphere are suspect. In fact, the vertical profile of upper troposphere humidity from the model is consistent with the satellite observations, which are only valid for large clear-sky regions. This implies that the model may not be an accurate representation of the humidity structure of the whole tropical atmosphere including both cloudy and clear-sky regions.

In order to examine the connection between convection, humidity, and the greenhouse effect experimentally, extensive calibrated measurements of water vapor were collected during CEPEX. We present preliminary results of the vertical structure of water vapor along the 2° S meridian, between Honiara (Solomon Islands), 160°E and Christmas Island, 160°W. Because of the complex processes that couple water vapor and hydrometeors in clouds, the quantitative relation between the intensity of deep convection and the distribution of water vapor should be obtained from orthogonal measurements. The intensity of convection should be derived from observations that are independent of water in any of its aggregate states. *Smit et al.* [1990, 1993] have used tropospheric ozone to map deep convection. Their concept relies on the fact that ozone close to the sea surface is destroyed by the following photochemical reactions in the absence or near absence of NO_X:

$$O_3 + h\nu \rightarrow O(^1D) + O_2 \quad \text{(R1)}$$

$$O(^1D) + H_2O \rightarrow 2OH \quad \text{(R2)}$$

$$OH + CO + O_2 \rightarrow HO_2 + CO_2 \quad \text{(R3)}$$

$$O_3 + HO_2 \rightarrow OH + 2O_2 \quad \text{(R4)}$$

Low concentrations of NO_X are prevalent in the maritime intertropical convergence (ITCZ). Therefore the ITCZ, with its large water vapor concentration and net upward - directed flow of air, represents a flowing processor of tropospheric ozone. Any ozone near the surface that is transported into the ITCZ is destroyed within a few days. The ozone lifetime is much longer starting at 2–5 km above the surface because the water vapor mixing ratio decreases rapidly with altitude. Extremely low ozone concentrations in the middle and upper troposphere indicate that boundary layer air has been lifted through an environment with larger ozone concentrations. Thus the ozone can serve as a tracer of deep convection. It is the near absence and not the presence of ozone that is used to map the large scale structure of deep convection.

2.2 Experimental procedure

Water vapor and ozone soundings were made from the R/V *Vickers* during its cruise from Honiara to Christmas Island. Water vapor was measured every 6 hours with conventional Vaisala thin film (Humicap) capacitive sensors (the MPI instrument series). On a second set of radiosondes, water vapor was measured with Humicap sensors and ozone concentration was measured with electrochemical ECC cells (the NOAA/KFA instrument series). A NOAA frost point hygrometer was substituted for the Humicap water vapor device on approximately half of the ozone soundings. The hygrometer has been used extensively to test satellite instruments designed to measure water vapor in the stratosphere. During CEPEX, the hygrometer served as a reference instrument to assess the accuracy of the Humicap results above 5 km under conditions of active and suppressed convection.

2.3 Experimental results

Preliminary comparison of the frostpoint hygrometer and Humicap observations indicates that the water vapor data from the two instruments are in good agreement between 5 and 13 km. The relative humidity (over ice for $T < 0$ C) is plotted versus longitude in Figure 5 (NOAA/KFA series). Figure 6 shows relative humidities as obtained from the MPI series. 18 Humicap water vapor soundings have been used to construct Figure 5 while 44 Humicap soundings are plotted in Figure 6. Therefore the zonal resolution of Figure 6 is better than that of Figure 5. These figures show that, to the west of the dateline (160° to 180°E), the troposphere was humid throughout the troposphere. Conversely, east of the dateline the troposphere above the first 2 km was dry up to about 13 km. Above 13 km, the humidity increased with altitude but did not reach saturation. Comparison of Figures 5 and 6 reveals slight differences which, at this preliminary stage of the analysis, are not considered significant. (It is likely that the humidities near the surface from the MPI series will increase somewhat as the data reduction progresses).

The vertical cross section of ozone with longitude is shown in Figure 7. The mixing ratios at the surface are very low. West of the dateline, the very low mixing ratios are confined to a shallow layer of less than 0.5 km above surface, while east of the dateline the mixing ratios are extremely low within one km or so above surface. Between 180° and approximately 172°W, larger mixing ratios (approximately 15 to 20 ppb) extend from the middle troposphere to the tropopause. For most of the cruise the upper troposphere was nearly devoid of ozone. This is particularly apparent between 160°E and 180°E. The measurements are upper limits in this region because the finite value of the ECC background current and incomplete knowledge of its dependence on altitude did not allow more accurate measurements. A vertical profile of ozone and relative humidity for the first sounding, at 160°E, is shown in Figure 8. This sounding clearly shows the pronounced anti-correlation of ozone and water vapor at 3 to 4 km.

2.4 Analysis

When the concentration of NO_X is low near the surface, ozone is depleted in only 4 days by the photochemical reactions R1 to R4 [*Smit et al.,* 1993]. The 850 mb analysis [*Ramanathan et al.,* 1993] shows long fetches of easterly flow, almost parallel to the equator. Ozone cannot be transported close to the surface over long distances, since it would be destroyed in a matter of a few days. In the zonal plane of Figure 8, the source of ozone detected near the surface is the middle troposphere, which seems to be supplied with ozone from the stratosphere at longitudes close to the dateline. However, a complete meteorological analyses of the large scale flow is required to determine the three-dimensional transport of ozone between the stratosphere and troposphere. It is clear, however, that the source of ozone in the middle troposphere is related to atmospheric subsidence because of the pronounced anti-correlation between ozone and water vapor. The potential temperature at which the anti-correlation was observed is 315 to 320 K, and the corresponding isentropic surface is connected to the stratosphere at extratropical latitudes. H_2O/O_3 profiles similar to that of Figure 8 were previously found over the ITCZ of the Atlantic ocean under conditions of deep convection [*Smit et al.,* 1990].

Vertical H_2O/O_3 profiles such as depicted in Figure 8, and the large scale distribution to the west of the dateline shown in Figures 5 to 7, suggest that the water vapor budget for the areas with deep convection has two components. The first component is the vapor lifted vertically from a shallow layer between sea surface and a height of about 3 to 4 km, equivalent to a potential temperature of about 300 to 320 K. The relative humidity of this air is large, approximately 80%. The second component is the quasi-horizontal advection of dry air that is mixed with the air lifted up from the boundary layer. The dry air may originate at extratropical latitudes. The resulting water vapor mixing ratio of the air that is transported to the upper troposphere can be constrained by the observed ozone mixing ratios. For the western part of the CEPEX cruise the ozone mixing ratios in the upper troposphere are as low, maybe even lower, than the mixing ratio at sea surface. This would imply that little entrainment took place. Conversely, *Smit et al.* [1993] found that the entrainment was larger over the equatorial Atlantic

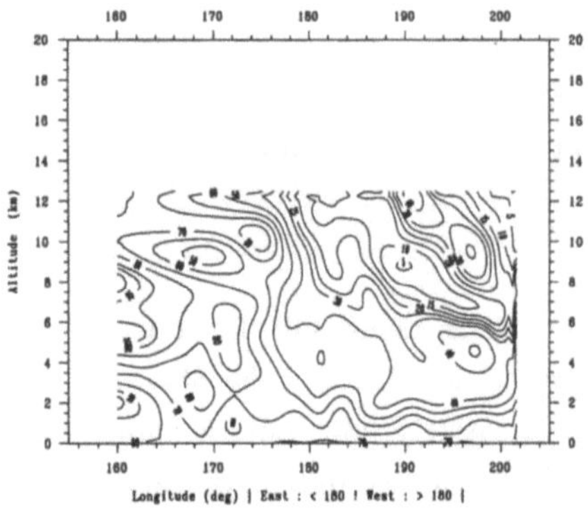

Figure 5: The height versus longitude cross section of relative humidity (over ice for $T < 0$ C) from soundings launched from the R/V *Vickers* between Honiara and Christmas Island (NOAA/KFA instrument series).

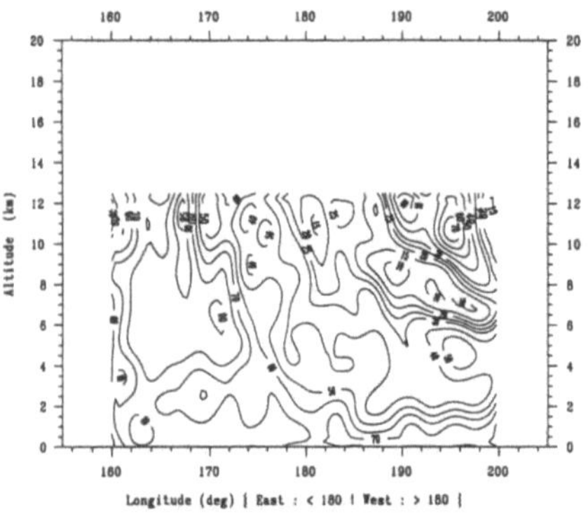

Figure 6: The height versus longitude cross section of relative humidity (over ice for $T < 0$ C) from soundings launched from the R/V *Vickers* between Honiara and Christmas Island (MPI instrument series).

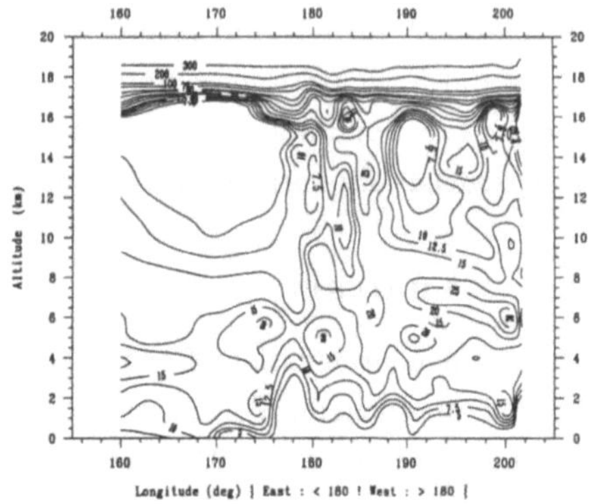

Figure 7: The height versus longitude cross section of tropospheric ozone corresponding to Figure 5 (NOAA/KFA instrument series).

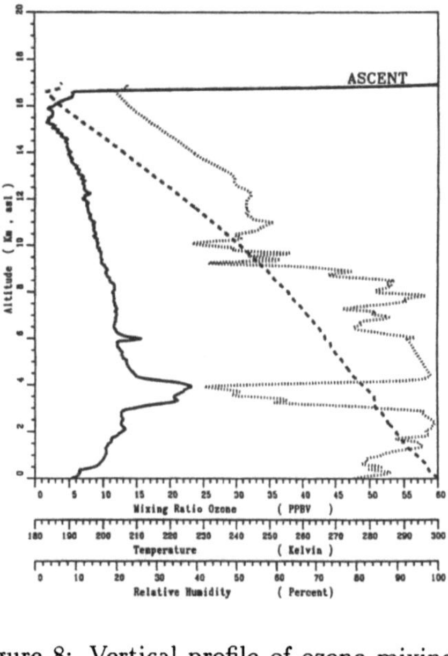

Figure 8: Vertical profile of ozone mixing ratio (straight line), relative humidity (dotted line), and temperature (dashed line) for the first sounding, which was launched at 160°E close to Honiara.

ocean. Since the ozone mixing ratio of the upper troposphere is the weighted mean of the vertically lifted plus the quasi horizontally advected flow, we may be able to deduce the water vapor budget of the outflow region by a measurement of its concentration. This should help determine the mesoscale effects of convection on the distribution of water vapor.

3 Cloud-radiation feedback and interactions

3.1 Introduction

The relation between tropical convection, the atmospheric greenhouse effect, and surface longwave cloud forcing has been investigated using the sounding data collected during CEPEX. The aim of this investigation is to estimate the magnitude of the longwave heating terms in the heat budget (eq. [2]) and the gradient of these terms with SST. RC found that much of the warmest tropical ocean is unstable with respect to cooling by longwave radiation. This instability is manifested in the clear-sky greenhouse effect for the surface-atmosphere column defined by:

$$G_a = \sigma T_s^4 - F_a^+(T) \tag{7}$$

$F_a^+(T)$ is the outgoing longwave flux at the top of the atmosphere. The increase in G_a with T_s is greater than the corresponding increase in surface emission. The onset of the supergreenhouse effect coincides with the threshold temperature for deep convection of 300 to 301 K [*Gadgil et al.,* 1984; *Graham and Barnett,* 1987]. It has been shown that the supergreenhouse effect requires an increase in absolute humidity with SST that is much greater than the rate predicted by the Clausius-Clapeyron relation with constant relative humidity [*Minschwaner and McElroy,* 1992; *Hallberg and Inamdar,* 1993]. One possible mechanism for moistening the atmosphere is vertical advection of water by deep convection. One of the main objectives of CEPEX was to determine the greenhouse heating both at the surface and the tropopause.

Because the upper tropical troposphere is much colder than the sea surface, optically thick cirrus anvils associated with deep convection can reduce the outgoing longwave radiation by 100 $\mathrm{W\,m^{-2}}$ or more. The partitioning of the column longwave heating by the cirrus between the ocean and the atmosphere is determined by the infrared optical opacity of the atmosphere and the variation of temperature with altitude. Radiative transfer calculations indicate that most of the heating from the cloud longwave absorption is confined to the middle and upper troposphere [*Slingo and Slingo,* 1988; *Harshvardhan et al.*, 1990]. Therefore the heating effect of the upper tropospheric clouds should be a relatively small component of the surface energy budget H (eq. [2]). The decoupling of the surface from the tropospheric cloud forcing [*Stephens and Webster,* 1984] is a critical component of the cirrus thermostat mechanism. The decoupling was parameterized by RC using the ratio of the surface to the top-of-atmosphere longwave cloud forcing:

$$f = \frac{C_l^*}{C_l(T)} \tag{8}$$

Another principal objective of CEPEX was to measure the cloud radiative forcing at the surface, including the longwave f factor.

3.2 Methods and data

The clear-sky longwave fluxes in the atmosphere have been calculated from the Vaisala water vapor soundings described in section 2 using a version of the LOWTRAN 7 radiative transfer code [*Kneizys et al.,* 1988]. The model includes contributions from H_2O (including temperature-dependent continuum absorption), CO_2, O_3, CH_4, and NO. The trace gas data are from the McClatchey standard tropical atmosphere. Two additional sets of soundings have been constructed. In the first set, the SSTs and vertical temperature profiles are replaced by the mean temperature profile. In the second set, the vertical profiles of water-vapor mixing ratios are replaced by the average mixing ratio profile. The standard, fixed temperature, and fixed water vapor profiles are used to identify contributions to the gradients in G_a from gradients in the water vapor and lapse rate. The fluxes calculated from soundings launched in cloudy conditions give the effect of clouds on the surrounding air.

The surface cloud forcing C_l^* has been derived using measurements from a Fourier Transform Infrared spectroradiometer (FTIR) on the *Vickers.* The FTIR measures the downwelling radiances in the mid-infrared spectral range with a detector aligned along the local zenith. The integrated downwelling flux in this spectral interval is estimated using limb-darkening models derived from the Vaisala soundings. Infrared absorption bands outside the spectral bandpass of the FTIR are virtually saturated in the tropical atmosphere, and therefore do not contribute to C_l^*. The cloud forcing at the top of the atmosphere, $C_l(T)$, has been derived from the narrowband mid-IR radiometer onboard the Japanese GMS-4 Geostationary Meteorological Satellite. The satellite recorded the outgoing radiance field with 5 km resolution in the neighborhood of the ship every hour. The relationship of the outgoing narrowband radiances and longwave fluxes has been calculated using a delta-Eddington radiative transfer code with realistic cloud properties.

This relationship is used to convert the GMS-4 observations to estimates of the longwave cloud forcing.

3.3 Greenhouse and cloud radiative heating of the surface

The time series of the column greenhouse effect are plotted in Figure 9 using the standard, fixed temperature, and fixed water vapor profiles. Satellite observations of outgoing longwave radiation (OLR) indicate that deep convection was prevalent west of the dateline during the first four days of the voyage (March 8 to 11, 1993). From March 12 to 16, the Vickers observed relatively clear conditions as its sailed from from 178° to 196°E. Between March 17 and 21, the convective activity near the ship increased. The time series of G_a shows that G_a is larger during the convective periods. The time series of G_a calculated from the fixed temperature profiles is well correlated with the G_a calculated with the standard set of soundings. The gradient in G_a from the fixed temperature profiles is 86% of the total gradient in G_a. In addition, the time series of G_a from the fixed vapor soundings is poorly correlated with the G_a from the standard soundings. This demonstrates that most of the variability and the east-west gradient in G_a is due to changes in water vapor. The gradient in the greenhouse trapping in the upper troposphere between 200 and 500 mb is 7.2 $\mathrm{W\,m^{-2}}$, approximately 41% of the gradient for the entire column. Therefore one of the main factors contributing to the rapid rise in G_a from clear to cloudy regions is a significant moistening of the upper troposphere.

The variation of G_a with SST and an index of convective activity is shown in Figure 10. The convective index is the infrared brightness temperature (IRBT) measured by the GMS-4 in a 100 km^2 neighborhood of the ship. For optically thick clouds in the middle and upper troposphere, the IRBT are approximately equal to the cloud top temperatures. Regions of clear sky have IRBT that exceed 280 K. In clear regions with IRBT greater than 280 K, the variation of G_a with SST is consistent with aircraft observations during CEPEX [*Valero et al.*, 1993]. By recalculating the phase diagram with the fixed temperature and fixed water vapor profiles, we find that the effects of changing SST and water vapor are approximately equal in clear-sky conditions. However, the variation along the brightness temperature axis, in particular for fixed SST between 302 and 302.5 K, is almost entirely due to an increase in water vapor in convective regimes. The variation of G_a with IRBT is comparable to the variation of G_a with SST.

The net clear-sky longwave flux calculated from the soundings is shown in Figure 11. The net flux from the surface is given by $G_a^* - \sigma T_s^4$. The correlation between the net flux from fixed temperature and standard soundings indicates that changes in water vapor

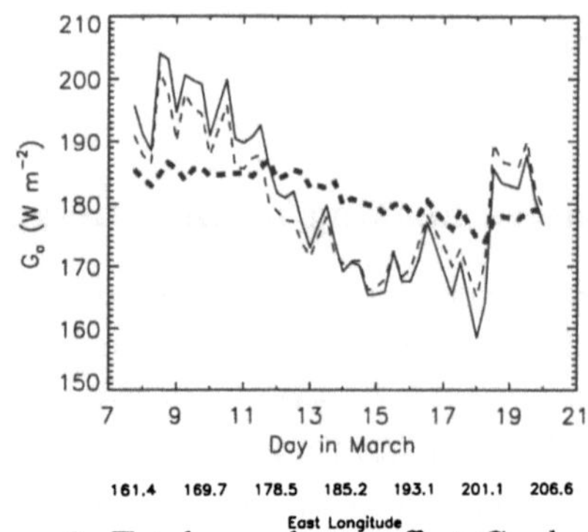

Figure 9: Total greenhouse effect G_a along the track of the R/V *Vickers* calculated from eq. (7) in $\mathrm{W\,m^{-2}}$. Solid line: standard soundings; thin dashed curve: fixed temperature soundings; and thick dashed curve: fixed water vapor soundings.

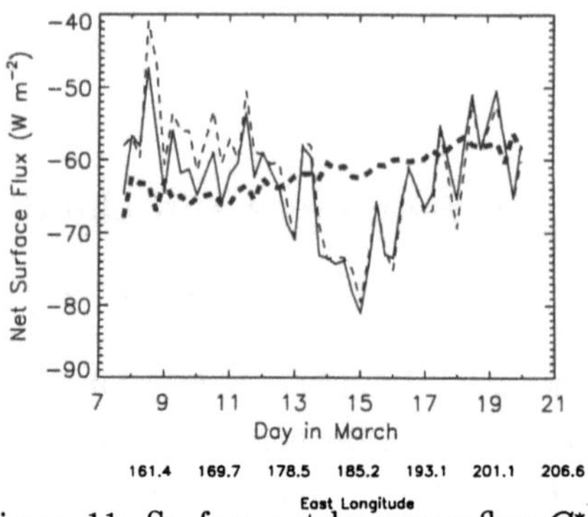

Figure 11: Surface net longwave flux $G_a^* - \sigma T_s^4$ along the ship track calculated from the soundings. Solid line: standard soundings; thin dashed curve: fixed temperature soundings; and thick dashed curve: fixed water vapor soundings.

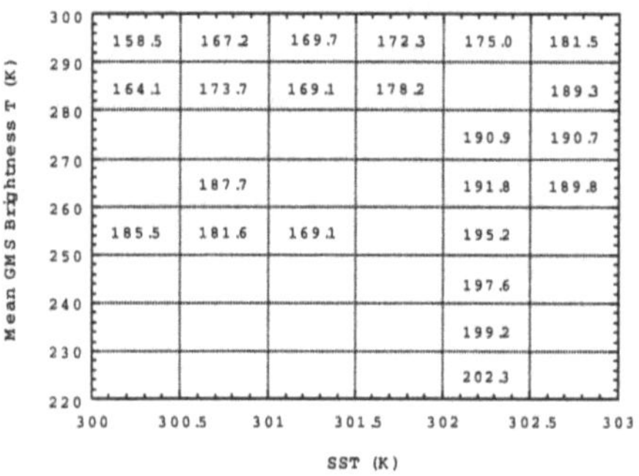

Figure 10: Phase diagram of G_a calculated from the soundings as a function of SST and mean infrared brightness temperature from the GMS weather satellite.

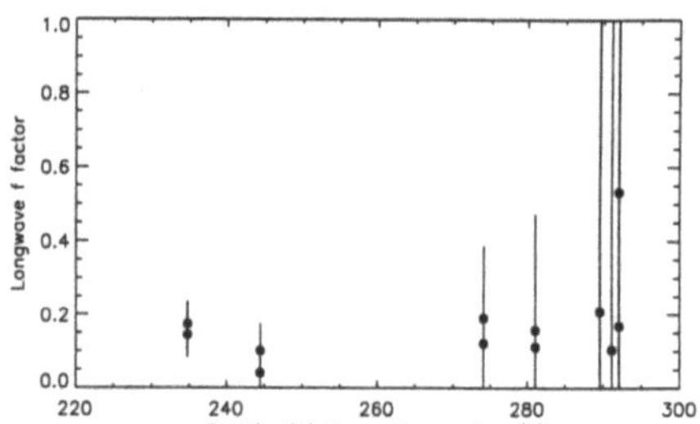

Figure 12: Longwave f factor (8) computed from collocated FTIR and GMS observations of cirrus clouds as a function of GMS infrared brightness temperature.

amount are responsible for most of the variability in the downward flux at the surface. The sharp maximum in the net flux magnitude on March 15 corresponds to minima in the column water vapor in the lower and middle troposphere. The minima in the net flux magnitude correspond to the periods of increased convection at the beginning and end of the *Vickers* transect across the central Pacific. These results confirm the direct measurement of the greenhouse effect at the surface by *Lubin* [1994] and show the connection between longwave emission from the atmosphere, convection, and the vertical distribution of water vapor.

The longwave f factor for cirrus clouds is plotted as a function of the satellite brightness temperature in Figure 12. The range of brightness temperatures is related to primarily to the variations in optical thickness of the cirrus. When the clouds are optically thin, the radiance measured by the satellite includes contributions both from the cold cloud and the warm ocean surface. The FTIR observations show that less than 20% ($f \leq 0.2$) of the infrared energy absorbed in an atmospheric column containing cloud is transmitted to the surface. This is consistent with estimates of the f factor used in the original formulation of the cirrus thermostat hypothesis.

3.4 Conclusions on greenhouse and cloud heating

In the CEPEX sample, when the SST reaches the threshold temperature for deep convection, large changes in G_a and G_a^* at fixed SST are associated with synoptic-scale deep convective events. This finding suggests that the total gradient in G_a and G_a^* with SST depends in part on the statistical correlation between SST and the frequency of deep convection. The observations indicate that there is a supergreenhouse effect for both the surface and the atmospheric column. The convection also heats the surface through the longwave cloud forcing, but the forcing is less than 20% of the total longwave radiation absorbed by the cloudy atmosphere. Since the shortwave cloud forcing at the surface C_s^* is at least as large as the shortwave forcing at the top of the atmosphere $C_s(T)$, and since satellite observations indicate that $C_s(T)$ and $C_l(T)$ are similar in magnitude, the net radiative effect of convection is to cool the ocean surface.

4 Evaporation

The analysis of the evaporation data from the 40 Hz. aircraft measurements was not complete at the time of the workshop and could not be included in the book. However, the data have been subsequently processed and analyzed, and the results have been sent for publication [*Zhang and Grossman*, 1995].

5 Conclusions

The initial analysis of the soundings and other measurements at the surface indicate that convection moistens the entire tropospheric column and, as a direct result, enhances the greenhouse trapping in the tropical atmosphere. Thus ocean mixed layer in the warm pool cannot cool by thermal emission alone. In terms of the simple model of the mixed layer presented in section 1, the processes that can stabilize the mixed layer are related to the shortwave and nonradiative components of the heat budget. These processes are being actively investigated with measurements from CEPEX and other field experiments. The results from these investigations should improve our empirical understanding of the warm pool and our ability to model the interactions of the tropics with the rest of the climate system.

References

Gadgil, S., P. V. Joseph, and N. V. Joshi, Ocean-atmosphere coupling over monsoon regions, *Nature*, *312*, 141–143, 1984.

Graham, N. E., and T. P. Barnett, Sea surface temperature, surface wind divergence, and convection over tropical oceans, *Science*, *238*, 657–659, 1987.

Hallberg, R., and A. K. Inamdar, Observations of seasonal variations in atmospheric greenhouse trapping and its enhancement at high sea surface temperature, *J. Climate*, *6*, 920–931, 1993.

Harshvardhan, D. A. Randall, and D. A. Dazlich, Relationship between the longwave cloud radiative forcing at the surface and the top of the atmosphere, *J. Climate*, *3*, 1435–1443, 1990.

Kneizys, F. X., E. P. Shettle, L. W. Abreu, J. H. Chetwynd, G. P. Anderson, W. O. Gallery, J. E. A. Selby, and A. A. Clough, *Users Guide to Lowtran 7*. Air Force Geophysics Lab., 1988, Rep. AFGL-TR-88-0177.

Lubin, D., The role of the tropical super greenhouse effect in heating the ocean surface, *Science*, *265*, 224–227, 1994.

Minschwaner, K., and M. B. McElroy, A model for the energy budget of the atmosphere: Comparison with data from the Earth Radiation Budget Experiment, *Planet. Space Sci.*, *40*, 1237–1250, 1992.

Newell, R. E., Climate and the ocean, *Am. Sci.*, *67*, 405–416, 1979.

Niiler, P. P., and J. Stevenson, The heat budget of the tropical ocean warm-water pools, *J. Marine Research, 40 (Supplement)*, 465–480, 1982.

Ramanathan, V., and W. Collins, Thermodynamic regulation of ocean warming by cirrus clouds deduced from observations of the 1987 El Niño, *Nature*, *351*, 27–32, 1991.

Ramanathan, V., R. Dirks, R. Grossman, A. Heymsfield, J. Kuettner, and F. Valero, *Central Equatorial Pacific Experiment Design.* Center for Clouds, Chemistry, and Climate, University of California, San Diego, CA 92093, 1993.

Slingo, A., and J. M. Slingo, The response of a general circulation model to cloud longwave radiative forcing. Part I: Introduction and initial experiment, *Quart. J. R. Meteorol. Soc.*, *114*, 1027–1062, 1988.

Smit, H., S. Gilge, and D. Kley, The meridional distribution of ozone and water vapor over the Atlantic Ocean between 30S and 52N in September/October 1988, in *Physico-chemical behavior of atmospheric pollutants*, edited by G. Restelli and G. Angeletti, 736 pp., Kluwer Academic Publishers, Dordrecht, 1990.

Smit, H., D. Kley, H. Voemel, and S. Oltmans, Longitudinal vertical distribution of tropospheric ozone and its correlation with water vapor over the equatorial Pacific Ocean between 160E and 160W, *EOS*, *74*, 115, 1993.

Stephens, G. L., and P. J. Webster, Cloud decoupling of the surface and planetary radiative budgets, *J. Atmos. Sci.*, *41*, 681–686, 1984.

Sun, D. Z., and R. S. Lindzen, Distribution of tropical tropospheric water vapor, *J. Atmos. Sci.*, *50*, 1643–1660, 1993.

Valero, F. P. J., P. Pilewskie, and A. Bucholz, Determination of the clear-sky greenhouse trapping of infrared radiation from multi-level aircraft measurements over the central equatorial Pacific, *EOS*, *74*, 107, 1993.

Vonder Haar, T. H. 1986. Surface radiation budget observations and analysis. in *Surface radiation budget for climate applications*, 87–101, . NASA Ref. Pub. 1169.

Waliser, D. E., and N. E. Graham, Convective cloud systems and warm-pool sea surface temperatures - coupled interactions and self-regulation, *J. Geophys. Res*, *98*, 12881–12893, 1993.

Young, G. S., D. V. Ledvina, and C. W. Fairall, Influence of precipitating convection on the surface energy budget observed during a Tropical Ocean Global Atmosphere pilot cruise in the tropical western Pacific ocean, *J. Geophys. Res.*, *97*, 9595–9603, 1992.

Zhang, C., Large scale variability of atmospheric deep convection in relation to sea surface temperature in the tropics, *J. Climate*, *6*, 1898–1913, 1993.

Zhang, G. J., and R. L. Grossman. March, 1995. Surface evaporation during the Central Equatorial Pacific Experiment: A climate scale perspective. submitted to *J. Climate*.

SOME REMARKS ON MECHANISMS FOR THE REGULATION OF TROPICAL SEA SURFACE TEMPERATURE

R. T. Pierrehumbert
Department of Geophysical Sciences
University of Chicago
Chicago, IL 60637

1. Introduction

It hardly needs elaboration that the regulation of tropical sea surface temperature is a central problem in climate, with repercussions ranging from the paleoclimatic domain (e.g. climate of the last glacial maximum vs. the warm climates of the Eocene) through the El Niño fluctuations of the present climate, and beyond to projections of climatic impact of anthropogenic CO_2 increase. My object in this brief communication is to convey some of the substance of the spirited and stimulating exchange of ideas on the subject that took place at the NATO ARW to which this volume is devoted. The dialogue has by no means reached closure, and no doubt, the issues I discuss here will continue to be debated for some time to come. It would be safest to view this as my own ideosyncratic view on the subject, which may be wrong in some of its details, but hopefully wrong in an interesting way — or at least in a different way from previous works on the subject.

The central fact of life in the tropics is the destabilizing water vapor feedback, which makes it difficult for a sufficiently moist atmosphere to radiate away tropical levels of insolation locally, without running away to temperatures far in excess of the observed temperature. In its extreme form, the uncompensated feedback can lead to a "runaway greenhouse," anticipated theoretically by Kombayashi (1967) and Ingersoll (1969) and developed further in the intervening work summarized in Renno *et al.* (1994). In the following I will be principally concerned with major "zero order" effects, and especially with the problem of what counteracts the tropical water vapor feedback and brings the sea surface temperature (SST) down to the vicinity of its presently observed range. There are any number of processes (e.g. lapse rate alteration or trade boundary layer adjustments) which could affect the tropical surface temperature by 2-3°K and changes of this magnitude are unquestionably important with regard to climate during an ice age or in a doubled CO_2 world. If one is

NATO ASI Series, Vol. I 35
Clouds, Chemistry and Climate
Edited by Paul J. Crutzen and Veerabhadran Ramanathan

interested in that level of accuracy, the physics must be treated meticulously and in detail, and the broad-brush approach I adopt in the following will not suffice. To be fair, it is not clear that the current generation of general circulation models suffices for this level of accuracy either.

There are currently three principal mechanisms on the table with regard to tropical SST regulation. They are evaporative feedback (Newell 1979; see also the review given in Waliser and Graham 1993), cloud albedo feedback (the "thermostat hypothesis of Ramanathan and Collins 1991, hereafter RC), and production of dry air in "radiator fins" (Pierrehumbert 1995, with antecedents in Sun and Lindzen 1993). I will argue against the first two proposals, which predict an extraordinarily stable tropical climate; unlike Sherlock Holmes I make no claims as to the validity of the third by default. The radiator fin model still has some notable missing pieces. It succeeds in counteracting the destabilizing water vapor feedback in the warm pool region, but gives little reason to expect tropical climate to be especially resilient.

2. Basic physics governing the temperature of warm oceans

Consider a single column of the atmosphere-ocean system, with atmospheric temperature profile T(z) and ocean surface temperature T_S. The problem of determining the climate can be divided into two steps: (**PROBLEM A**) the determination of T_S *given* T(z), and (**PROBLEM B**) the determination of the T(z) that satisfies the top-of-atmosphere radiation balance and the constraints of radiative-convective dynamics. Problem B is not as underdetermined as it looks, because the form of T(z) is highly constrained. In the stratosphere, it is uniquely determined by radiative equilibrium and the atmospheric composition. In the troposphere, the situation is somewhat more complicated, since the profile is set by the mixing due to moist convection. Nonetheless, observations and sophisticated models (Sun and Lindzen 1993; Emanuel 1991) indicate that the tropics stays sufficiently close to the saturated moist adiabat for that to constitute a satisfactory "zero order" description. The importance of this is that T(z) is (approximately) constrained to a 1-parameter family of curves, so that specifying the temperature at any one level determines the atmospheric temperature at all levels.

For a warm, moist atmosphere such as prevails in the tropics, changing T_S while keeping T(z) fixed yields little change in the infrared radiation escaping the top of the atmosphere (the "OLR"). This is because such atmospheres are optically thick in the infra-red. In consequence, Problem A decouples from Problem B. One can first determine T(z), then find T_S *ex post facto* by consideration of the surface energy budget.

A major complication to this picture is the necessity of specifying a water vapor and cloud water profile when solving the radiative transfer problem to determine the OLR. These profiles typically depend on many things besides T(z) itself. The classical prescription for water vapor is to fix relative humidity, but other approaches are emerging (Sun and Lindzen 1993, Emanuel 1991). Representation of cloud effects in radiative-convective models is at an even earlier and more exploratory stage of development.

Whatever prescription is used, the aim is to identify a reference temperature T_{ref} characteristic of the atmosphere, determine the OLR as a function of T_{ref}, and then to determine T_{ref} by balancing OLR against the absorbed solar radiation. In its simplest incarnation, one adopts a constant temperature lapse rate Λ, and uses the radiation physics to determine an effective radiating level z_{rad}. The state of the climate is then determined by $S = \sigma T(z_{rad})^4$, where S is the absorbed solar radiation. The low level air temperature is then $T(0) = T(z_{rad}) + \Lambda z_{rad}$. In Section 3, I will argue that evaporation makes the surface budget "stiff" over moist surfaces, so that T_s cannot deviate much from T(0). It thus emerges that the top-of-atmosphere radiation budget together with the lapse rate are the main players in determining the surface temperature.

The primacy of the top-of-atmosphere considerations is important to keep in mind when assessing cloud impacts, because — in the present climate — the cloud albedo and cloud greenhouse effects very nearly cancel each other at top-of-atmosphere in the tropics (RC). A complication is that the atmosphere is not in local radiative equilibrium in the tropics, so that one must allow for lateral heat transports into and out of the atmosphere-ocean column. In Section 3, I will summarize some arguments showing that, in equilibrium, these transports do not substantially alter the naive expectation that cloud effects approximately drop out of the determination of atmospheric temperature.

3. Some illustrative models

First let's consider the sensitivity of SST to removal of clouds and to changes in the oceanic heat flux *assuming that the atmospheric temperature remains fixed during these changes.* This is a rather unrealistic thought experiment, in that climate change will generally involve changes in the atmospheric temperature T(z), rather than just changes in surface temperature. However, there is much to be learned about the interplay of clouds and evaporation from this restricted problem; later, I will bring atmospheric temperature changes back into the picture.

The temperature fluctuations of a mixed layer ocean with mixed layer depth h are governed by

$$\frac{dT_s}{dt} = F_{surf} / (\rho_w C_{pw} h) \tag{3.1}$$

where T_s is the surface temperature, ρ_w is the density of sea water, and C_{pw} is the specific heat of the sea water. The surface layer heating is given by

$$F_{surf} = \alpha S_c + \gamma C_s + \{I^-(0) - \sigma T_s^4\} - E - F_{sens} + F_o \tag{3.2}$$

S_c and C_s are the clear-sky absorbed solar flux and the cloud shortwave forcing measured at the top of the atmosphere, as defined by ERBE (cf Ramanathan *et al* 1989). Further, α is a clear sky shortwave absorption coefficient representing the portion of the clear sky top-of-atmosphere solar flux which reaches the ground without being absorbed by the intervening atmosphere, and γ is a coefficient measuring the effects of cloud shortwave absorption on surface insolation. $I^-(0)$ is the downwelling infrared flux impinging on the surface, E is the evaporative heat flux, F_{sens} is the sensible heat flux, and F_o is the heat added by oceanic transports into the surface layer (negative values signify cooling by importation of cold water, from the sides or from below the mixed layer).

Let's represent evaporation by the bulk aerodynamic formula (Peixoto and Oort 1992)

$$E = \rho L C_d u^* q^* \tag{3.3}$$

where ρ is the air density in the boundary layer, L is the latent heat of vaporization, C_d is the drag coefficient, u^* is the characteristic velocity fluctuation, and q^* is the characteristic scale of fluctuation of the water vapor mass mixing ratio. q^* can be estimated using

$$q^* = q_{sat}(T_s) - r\, q_{sat}(T(0)) \tag{3.4}$$

in which q_{sat} is the saturation mixing ratio and r is the typical relative humidity of air entrained into the boundary layer. Following RC, I empirically represent the cloud shortwave forcing as

$$C_s = -a(T_s - T_{crit}) \quad \text{for } T_s > T_{crit} \tag{3.5}$$

with a = 20 $W/m^2\ {}^oK^{-1}$. C_s is set to zero below the T_{crit} = 300°K convective threshold, and saturates at the temperature where all the solar radiation is blocked. To simplify the discussion, I shall ignore F_{sens}.

I will now discuss some transient adjustment problems carried out with the above system, with the atmospheric temperature held fixed. The fixed-atmosphere assumption is implemented by specifying T(0) constant at 300°K and $I^-(0)$ constant at 393W/m^2, which is roughly the back radiation that corresponds to T(0) with a relative humidity at 80%. The other fixed parameters are: h = 40m, S_c = 370 W/m^2, α = .8, u^* = 4 m/s, r = 80%, and C_d = .0015. Based on the growing evidence for anomalous cloud shortwave absorption, we set γ=1.5 (Ramanathan, *et al* 1995), though this value should at present be regarded as somewhat speculative. With these parameters, the SST equilibrates at 302.1°K if F_o is set to -30 W/m^2. The equilibrium evaporation for this case is 114 W/m^2, which is entirely consistent with the observed mean value in the Pacific warm pool (Ramanathan *et al* 1995) if one allows 10 W/m^2 for the neglected sensible heat flux.

In the first adjustment experiment, one begins with the equilibrated system and instantaneously removes the clouds[1] by setting a=0. The adjustment of T_s and the evaporation is shown in Figure 1. Note that even complete removal of the cooling effect of clouds results in only a 2.3°K rise in the SST. This is because the excess insolation caused by removal of the clouds is accomodated mostly by a 60 W/m^2 increase in the evaporation, which takes about 100 days to come into being. The rise in evaporation does not require a great increase in SST, because evaporation is such a steeply increasing function of the difference between the SST and the overlying air temperature.

---

[1] Properly speaking, one ought to reduce $I^-(0)$ somewhat when removing the clouds, but because of the optical thickness of a warm moist atmosphere, this effect is rather inconsequential in comparison with the effect of clouds on surface insolation. Our calculation slightly exaggerates the warming effect of removing clouds, and therefore, if anything, overestimates the role of clouds.

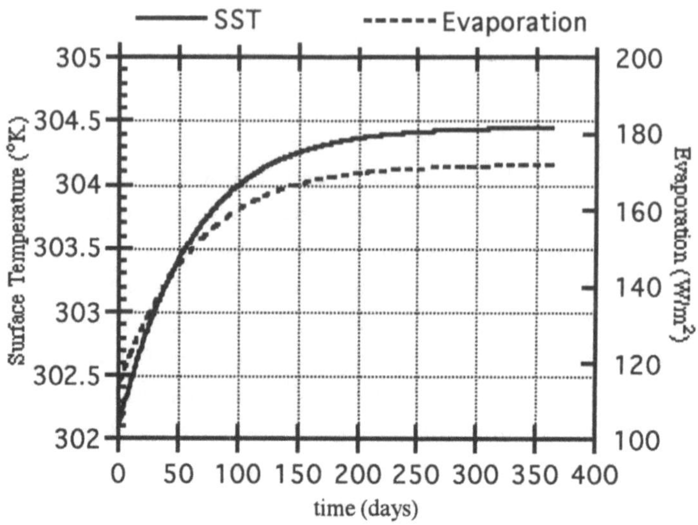

Figure 1. Adjustment of sea surface temperature and evaporation, following elimination of clouds. Calculation carried out with atmospheric temperature held fixed.

Clearly a 2.3°K change is not insignificant. In regions of light winds, clouds undoubtedly help hold down the air-sea temperature difference. My point is that the large effect of cloud shading on surface insolation can give an exaggerated notion of the sensitivity to clouds, if the buffering influence of evaporation is not taken into account. Modest increases in the wind speed or drag coefficient sharply reduce the impact of clouds. If we increase $C_d\, u^*$ from .006 to 0.01, the SST with clouds is 301.12, and without clouds is 302.02 — yielding a cloud effect of only .9°K.

To be sure, when clouds are eliminated, the evaporation rises to values far in excess of the long-term mean values that currently prevail over the warmest Pacific waters. This is no problem, because I am not claiming that there are no clouds in the Tropics — only that the air sea temperature difference wouldn't increase terribly much if clouds were eliminated. Moreover, the flatness of scatter plots of observed evaporation vs. SST (Zhang and McPhaden, 1995) poses no contradiction to the kind of evaporation model used in computing Figure 1. The analysis of Zhang and McPhaden implies only a *correlation* between high temperatures and the low windspeeds that limit evaporation. It does not imply that increasing the SST by any means and at any place will inevitably reduce winds. Given that the tropical winds are determined by aspects of the general circulation (like the Hadley cell) spanning the entire tropics, there are good reasons to doubt a local, causal connection. It seems likely that low winds cause high SST, rather than the other way around.

As an exercise in sensitivity, I next compare the adjustment of SST upon elimination of the oceanic cooling for the cloud and the no-cloud case. In both cases the SST is initialized with the equilibrium temperature for F_o = -30 W/m^2. Results are shown in Figure 2.

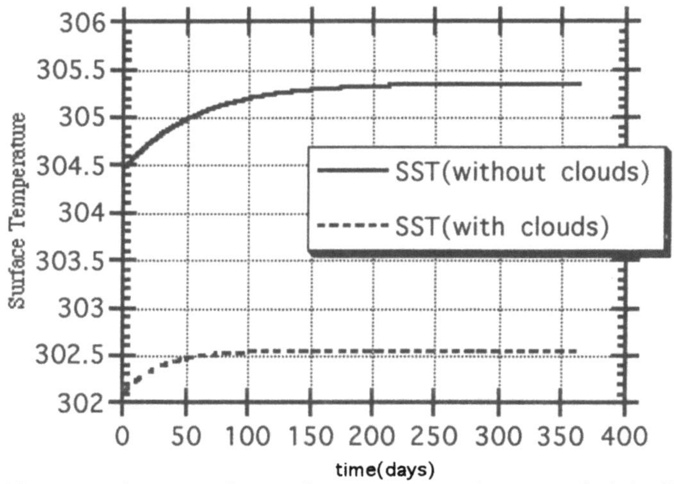

Figure 2. Adjustment of sea surface temperature after removal of the 30 W/m^2 oceanic cooling. Results are shown with and without clouds.

Because of the influence of evaporation, the difference in sensitivity between the two cases amounts to just .45°K.

From the preceding one concludes that evaporation has a powerful buffering effect on the surface energy budget, making the air-sea temperature difference resistant to large changes in the terms in the budget. Lindzen, Hou and Farrel (1982) also noted this effect. My argument is in essence the same given by Newell (1979), save that Newell overstated the implications of the result because he failed to take into account the fact that climate perturbations generally change T(0). *I am not arguing that evaporation exerts the primary control on the SST itself.* It keeps SST from much exceeding T(0), but does nothing to keep the atmospheric temperature from running away to large values. To determine T(0) itself, one must look to the top-of-atmosphere energy budget.

It might be thought that the preceding calculation is irrelevant to the debate on the thermostat hypothesis, because of the assumption of fixed atmospheric temperature. In fact it is very generous to the operation of a cloud thermostat , since the cloud longwave forcing C_l is prevented from warming the atmosphere, and from warming the surface. In reality, the atmosphere will warm "some" in response to C_l. How much? The answer to this question requires an estimate of the lateral heat flux in the atmosphere, which can carry away some of the excess heating of the atmospheric column and allow it to equilibrate at a lower temperature than would be the case for an isolated column.

The atmospheric heating which forces the atmospheric circulation can be written

$$Q_{atm} = \{\sigma T_s^4 - I^-(0)\} + (1-\alpha)S_c + (1-\gamma)C_s + Q_{L1} + C_l - I_c^+(\infty) \quad (3.6)$$

where $I_c^+(\infty)$ is the clear-sky OLR extrapolated to the whole scene as defined by ERBE, Q_{L1} is the latent heating of the column *that would result from local evaporation from the ocean below the column,* and the rest of the symbols have the same meaning as before. Note that (3.6) does not contain terms for latent heating due to redistributed moisture, nor for adiabatic cooling due to mean ascent; these are considered as part of the lateral atmospheric heat transport, since they are tied to the circulation driven by the heating. The atmospheric equilibrium temperature is determined by balancing Q_{atm} against the atmospheric horizontal heat transport F_{ah}. The budget (3.6) involves C_l but does not appear to involve all of C_s , which appears explicitly only in the solar absorption term. However, the rest of C_s can work its way back into (3.6) in a sneaky way. The change in surface insolation implied by C_s results in a change in the terms σT_s^4 and Q_{L1} (which equals E in a convecting region) as the surface budget attempts to come into equilibrium. For example, the dominant balance in the surface budget is between insolation and evaporation, so that reductions in insolation caused by increasing C_s eventually lead to strong reductions in E and Q_{L1} (Washington and Meehl 1993, Boer 1993). It was shown in Pierrehumbert 1995 that regardless of the detailed form of the terms in the surface energy budget, the cloud effects show up in Q_{atm} only in the combination C_s+C_l, once the surface budget comes into equilibrium.

The preceding considerations apply even if the atmospheric transports are computed using a full GCM, but further insight can be gained using the observation that atmospheric heat transports in the tropics are effective enough to keep the atmospheric temperature nearly horizontally homogeneous. This means that T(z) is determined by an average of Q_{atm} over the whole tropics rather than by its local behavior (except for behavior in a shallow boundary layer, which can tolerate more geographical variation because of the weak pressure gradients induced by shallow temperature anomalies.) Thus, T(z) and the moisture profile need to be adusted so as to satisfy

$$\langle Q_{atm} \rangle = 0 \qquad (3.7)$$

apart from the effects of atmospheric heat export to midlatitudes. The implied lateral heat flux within the tropics is $F_{ah} = \langle Q_{atm} \rangle - Q_{atm}$, and thus depends on all the heat sources enumerated in (3.6), and also on the behavior of the terms over the whole tropics, not just in the warm pool region. In contrast RC, who also invoke the uniform temperature approximation, argue that $F_{ah} = -(1\text{-}f)C_l$, for some small factor f; this estimate is sensitive only to the local cloud longwave heating of the atmosphere.

In order to illustrate some of the consequences of these ideas, I will now extend the earlier SST adjustment model to allow for changes in the atmospheric temperature. In this model I will make some simplifying assumptions which are not quantitatively accurate, but which enable us to focus on the main players while retaining physical consistency. First, I assume that the atmosphere is transparent in the visible (α=1, γ=1) and that the net IR cooling of the surface $\{\sigma T_s^4 - I^-(0)\}$ is negligable; corresponding terms are also dropped from the surface budget, to preserve consistency. To evaluate (3.7) the system is expanded about a presumed equilibrium with T(0) = 300°K, so that the clear sky OLR term is approximated by

$$\langle I_c^+(\infty) \rangle = OLR_o + \beta_1 (T(0) - 300°K) \qquad (3.8a)$$

where β_1 is some constant characteristic of the moisture distribution over the whole tropics. The "warm pool" region is assumed to occupy a fractional area $A<1$ of the tropics, and the changes in Q_{L1} relative to the base state are assumed to vanish outside this region. Similarly, the cloud effects are limited to the warm pool region. With these assumptions, (3.6,3.7) can be expanded around equilibrium to yield

$$0 = \langle Q_{atm} \rangle = -\beta_1 (T(0) - 300^oK) + A\{E - E_o + C_l - C_{lo}\} \qquad (3.8b)$$

where E_o and C_{lo} are the equilibrium evaporation and cloud longwave forcing corresponding to $T(0) = 300^oK$. The evaporation and cloud forcing have been assumed horizontally uniform over the warm pool region. E and C_l are determined as in the previous SST adjustment model, with the further assumption that $C_s(T_s)+C_l(T_s) = 0$. The surface temperature is determined as before by integrating (3.1) with the appropriate terms dropped from (3.2). With (3.8b), the changes in T(0) needed to keep the atmosphere in equilibrium in the face of changing E and C_l can be found; effects due to the finite adjustment time of the atmosphere are ignored, though some calculations (Emanuel 1991) suggest that the water vapor content of the atmosphere can take quite a long time to come into equilibrium. The calculations below were carried out with $\beta_1 = 2\ W/(^oK\,m^2)$. More comprehensive models of the equilibrium state without so many limiting assumptions, and including an explicit computation of β_1 in terms of a radiative model of the dry, subsiding regions of the tropics, are given in Pierrehumbert 1995.

In the first experiment, I take a system which is initially in equilibrium with $F_{oh} = -30\ W/m^2$, and then instantaneously eliminate the clouds. The subsequent adjustment of the atmospheric temperature is shown below, for two different choices of the warm-pool area.

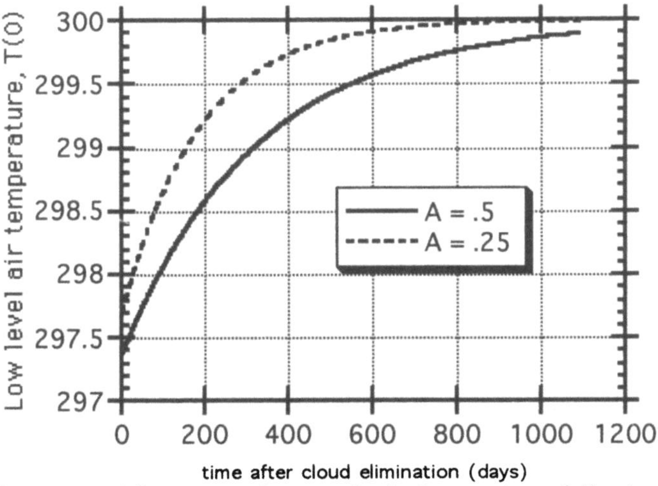

Figure 3. Adjustment of atmospheric temperature following instantaneous elimination of clouds.

Initially, the atmospheric temperature *drops* several degrees. This is because the cloud longwave heating C_l has been removed from Q_{atm}, but the evaporation has not yet had time to adjust to bring the surface budget back into equilibrium. Because a heating term has been removed, the atmospheric temperature must go down so as to reduce the OLR and bring the atmospheric budget into balance. As time goes on, however, the surface comes back into equilibrium, the evaporation rises to compensate the elimination of C_s from the surface budget, and the resulting change exactly compensates for the elimination of C_l — supposing, as always, that $C_s+C_l = 0$. In the end, the atmosphere returns to the same 300°K equilibrium temperature it had in the presence of clouds, though it requires a year or so for the recovery to take place.

In the next thought experiment, I consider the adjustment of the atmosphere to a new equilibrium following elimination of the 30 W/m^2 oceanic cooling in the warm pool region alone. The calculation is performed with and without clouds, but in each case the surface temperature is initialized in equilibrium with the 30 W/m^2 cooling prior to its elimination. Hence the atmosphere is initially in equilibrium at its unperturbed temperature of 300°K. Results are shown in the following graph.

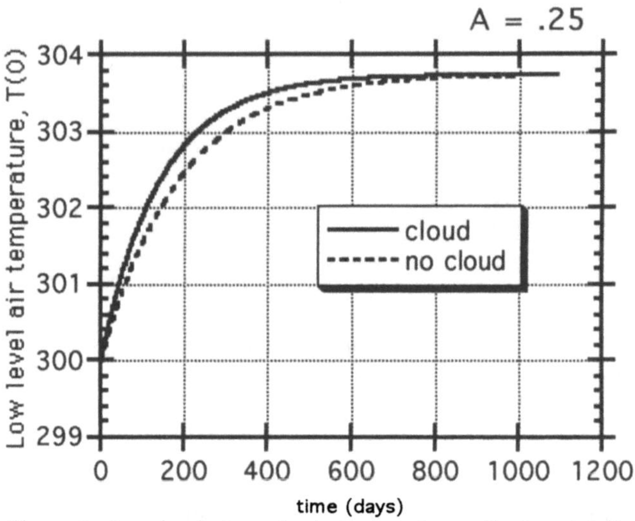

Figure 4. Low level atmospheric temperature adjustment following elimination of 30 W/m^2 oceanic cooling., with and without cloud feedback. Both calculations were initialized with the equilibrium corresponding to the 30 W/m^2 cooling. The convective region was assumed to occupy 25% of the area of the tropics.

Clouds affect the rate of approach to the new equilibrium, but eventually the surface budget comes back into balance so that clouds have no effect on the atmospheric temperature that is ultimately attained. In both cases, the atmospheric warming caused by elimination of F_o is 3.75 °K, which is fixed by averaging over the whole tropics the 30 W/m^2 change in warm-pool oceanic cooling (yielding 7.5 W/m^2 for A = .25), and dividing by the clear-sky OLR coefficient β_1.

In equilibrium, clouds thus have little impact on the atmospheric temperature, insofar as $C_s+C_l = 0$. If clouds are instantaneously removed from a system in equilibrium, elimination of clouds actually *cools* the atmosphere, during the time the surface budget is out of equilbrium. Conversely, sudden introduction of clouds to a cloud-free atmosphere would temporarily *warm* the atmosphere, making it harder for any cloud thermostat to prevail. If lateral heat transports are effective enough, the atmosphere may not warm *much*, but if the atmospheric warming becomes negligable we are back in the domain of the fixed-atmosphere calculation.

The sensitivity of the warm pool sea surface temperature to perturbations can be summed up in the sensitivity coefficient[1] β, defined so that

$$\Delta T_s = \frac{\Delta F}{\beta} \qquad (3.9a)$$

where ΔF is the imposed alteration of the energy flux into the atmosphere-ocean column and ΔT_s is the shift of warm pool SST in response. For example ΔF could be the roughly 4 W/m^2 radiative forcing due to doubling CO_2, or the roughly 25 W/m^2 shift in oceanic heat flux that might be expected in the course of El Niño fluctuations. In Pierrehumbert (1995) it is argued by systematic application of surface equilibrium, the uniform-temperature approximation, and the C_s+C_l cancellation that

$$\beta = \frac{\partial <I_c^+(\infty)>}{\partial T} \qquad (3.9b)$$

where T represents the atmospheric temperature, $I_c^+(\infty)$ is the *clear sky* outgoing longwave radiation as defined in ERBE, and angle brackets denote a horizontal average over the tropics. This stability estimate ignores the possibility of stabilizing or destabilizing feedbacks due to systematic shifts of heat exports from the tropics to the extratropics, and also neglects the modest effects of clouds on the sensitivity of the air-sea temperature difference. The two-box model of the tropics discussed in Pierrehumbert (1995) yields $\beta \approx$ 2 $W/(^oKm^2)$. Owing principally to their different assumption about lateral atmospheric heat transport, RC's estimate of sensitivity is dominated by the $\partial C_s/\partial T_s$ term, and yields sensitivity coefficients upwards of 15 $W/(^oKm^2)$, suggesting an extraordinarily stable tropical climate. Clearly, one of these estimates is seriously in error.

---

[1] There is a difference in sign between the definition of β employed in RC and that used in Pierrehumbert (1994). Here, we adopt RC's sign convention.

4. Prospects for clouds as a thermostat?

An issue of general importance to the effect of clouds on climate is the dependence of cloudiness on atmospheric temperature. To be sure, ERBE results (see esp. RC) show a dramatic increase in tropical cloud forcing as the SST exceeds a trigger temperature of about 300°K. However, this critical temperature does not represent an eternal thermodynamic constant for the planet, but rather the temperature necessary for the onset of deep convection. Deep convection is possible when the moist static energy of the low level air imparts sufficient buoyancy for air parcels to reach the upper troposphere, and so the critical temperature would increase if the atmosphere became warmer. With a warmer atmosphere a cloud thermostat, if any, would not even begin to kick in until higher temperatures, and so it is hard to see how clouds could exert a fundamental control on the atmospheric temperature. A nice illustration of this effect is provided by the GCM studies of Washington and Meehl (1993), carried out with a mixed layer ocean. Owing to the neglect of oceanic heat transports, their model equilibrates at a warmer tropical temperature than is the case for the real atmosphere; in accord with this, the threshold for onset of deep convection (as measured by the sharp rise of the atmospheric greenhouse effect) is 304°K rather than 300°K. Thus, the use of a fixed cloud-SST relation in my models of atmospheric adjustment discussed in the previous section is not strictly consistent, and may exaggerate the effects of clouds on the transient behavior.

Do we expect a generally warmer climate to be more or less cloudy than the present one? Given sufficient moisture supply one could argue (though not necessarily correctly) that a warmer atmosphere would be generally moister, owing to the Clausius-Clapeyron relation. There are no such simple limits for condensed cloud water, however. Cloudiness depends on the production of saturated air, not on the water vapor content of air. The condensate burden is limited by settling, evaporation into unsaturated air, and coalescence/precipitation. None of these effects is primarily controlled by temperature. In fact, if warmer air were generally more cloudy, we would expect the warm low level air to be densely swathed in clouds. Cirrus clouds would be practically nonexistent.

Dynamics as well as microphysics plays a crucial role in determining the the source of high altitude clouds. Dynamical transports keep T(z) relatively uniform over the whole tropics, which means that it cannot warm locally to accomodate local anomalously warm SST's. This process keeps T(z) cool enough to permit vigorous, deep convection in localized convective regions, at the expense of suppressing convection elsewhere. This modulation is supported further by the direct effect of the ascending and subsiding branches of the Hadley/Walker circulation, which powerfully enhance or suppress convection in the conditionally unstable atmosphere. Very deep convection provides the water vapor injection needed to create a dense upper tropospheric cirrus shield. The cloud patterns would be very different in a world with more uniform tropical SST's, which would entail more ubiquitous though less intense convection.

I have been deliberately vague in using the term "cloudiness" above. In reality, it is not just cloudiness, but the imbalance between cloud longwave and shortwave forcings that count. What becomes of the the approximate cancellation between C_s and C_l when we go to a generally warmer (2xCO_2, Eocene or Cretaceous) or cooler (ice age) atmosphere? This stands as the most pressing unresolved question concerning cloud effects on climate.

5. Conclusions

According to my picture, the cloud thermostat hypothesis, as stated in RC has the following three shortcomings:

- Viewed from the standpoint of the surface energy budget, it doesn't take into account the large evaporation term, which can change and "buffer" changes in cloud shortwave forcing.

- From the standpoint of the atmosphere-ocean column budget, RC's estimate of lateral atmospheric heat flux errs in not properly taking into account all heatings which drive the atmospheric circulation, and in not taking into account the importance of heating *contrasts* across the tropics. If there is not enough cooling somewhere in the tropics, the excess heating in the warm pool region cannot be gotten rid of, no matter how effective the lateral heat transport, and the tropics as a whole would have to warm significantly. The chief problem, though, centers on the behavior of latent heating in the warm pool atmosphere. In studying this issue through observations or GCM simulations, one must take care to distinguish transient effects from equilibrium behavior, since the latent heating adjustment will not be instantaneous.

- The critical surface temperature above which deep convection and tropical cloudiness increase is not a universal constant, but instead is expected on elementary physical grounds to increase or decrease in step with the overall atmospheric temperature of the tropics. Therefore, it is a fallacy to extrapolate to perturbed climates using a fixed critical temperature.

I am not claiming that clouds can never, under any circumstance, have a palpable effect on the tropical climate. Even if cloud shortwave and longwave forcing cancel at the top of the atmosphere, clouds could perhaps exert an influence through changing the vertical structure of the heating, through changing the relative humidity of the atmosphere, through changing the lapse rate, or perhaps through changing the transient response of the atmosphere-ocean system. These effects are subtle, and it is not even clear whether they are stabilizing or destabilizing. Then, too, the cloud longwave and shortwave forcing do not cancel exactly. Though the residual is small, it is nonetheless comparable to other small forcings driving the climate, such as the radiative perturbation due to doubling of CO_2. I would hardly wish to argue against the need for a better understanding of cloud effects.

Based on the above considerations, I would like to propose a "modified thermostat hypothesis." It is a considerably weaker and less dramatic assertion that the original one contained in RC, but it is what I think is actually supported by the CEPEX and ERBE data. The **Modified Thermostat Hypothesis** would read something like the following: "Insofar as surface winds tend to be weak over the warmest parts of the tropical ocean, evaporation alone cannot account for the tight coupling between SST and the overlying air temperature. In these regions, enhanced cloudiness plays a critical role in limiting the air-sea temperature difference." Without clouds, the air sea temperature difference in the warm pool region might be $2^{o}K$ larger than the observed value prevailing in the current climate.

Evaporation and clouds may jointly control the air-sea temperature *difference*, but this does not provide a fundamental limit to the tropical surface temperature itself, which is determined primarily by top-of-atmosphere considerations. While the true extent to which tropical SST has varied over geological time is still an unsettled issue, there are indications that the variations have been quite limited, particularly on the warm side. If these indications hold up, what could account for the apparent stability? I think the most likely candidate is the propensity of the large scale tropical circulation to create extensive pools of very dry air in undisturbed subsiding regions. These "radiator fins" are coupled dynamically to the rest of the tropics, and allow excess insolation to escape to space. Elsewhere (Pierrehumbert 1995) I have attempted to study their impact quantitatively, but much remains to be done before this problem can be laid to rest.

In my opinion, the original conception of CEPEX as a test of the thermostat hypothesis of RC has outlived its usefulness, in light of what has been learned by further thinking about the problem of SST regulation. This doesn't in any way compromise the importance of the observations taken during CEPEX. In fact, the outstanding success of CEPEX stands as a vindication of the concept of hypothesis-driven field experiments. The CEPEX data has already proved its worth in addressing questions related to cirrus lifetime and optical properties, and the factors governing the moisture distribution of the tropics. By any reckoning, such questions must figure prominently in any treatment of the effect of clouds in an altered climate.

Acknowledgements

This work was supported by the National Science Foundation, under grant ATM 89-20119 to the Center for Clouds, Chemistry and Climate.

References

Boer, G.J. 1993: Climate change and the regulation of surface moisture and energy budgets. *Climate Dynamics* **8**,225-239.

Emanuel, K. 1991: A scheme for representing cumulus convection in large scale models. *J. Atmos. Sci* **48,** 2313 - 2333.

Ingersoll, A. P., 1969: The runaway greenhouse: A history of water on Venus. *J. Atmos. Sci* **26,** 1191-1198.

Kombayashi, M. 1967: Discrete equilibrium temperatures of a hypothetical planet with the atmosphere and the hydrosphere of one component-two phase system under constant solar radiation. *J. Meteor. Soc. Japan*,**45,** 137-138.

Lindzen, R. S. , Hou, A. Y. and Farrel, B. F. 1982: The role of convective model choice in calculating the climate impact of doubling CO_2 *J. Atmos. Sci* **39,** 1189 - 1205.

Newell, R. E. 1979: Climate and the ocean, *Am. Sci.*,**67** 405-416.

Peixoto, J. and Oort, A. 1992: *Physics of Climate.* New York:Amer. Inst. Physics, 520pp.

Pierrehumbert, R. T. 1995: Thermostats, Radiator Fins, and the Local Runaway Greenhouse. *J. Atmos. Sci.* **52,** 1784-1806.

Ramanathan, V., Cess, R.D., Harrison, E.F., Minnis, P., Barkstrom, B.R., Ahmad, E. and Hartman,D. 1989: Cloud-radiative forcing and the climate: Results from the Earth Radiation Budget Experiment. *Science* **243**,57-63.

Ramanathan, V. and Collins, W. 1991: Thermodynamic regulation of ocean warming by cirrus clouds deduced from observations of the 1987 El Niño. *Nature* **351,** 27-32.

Ramanathan V., Subasilar B., Zhang G.J., Conant W. , Cess R.D., Kiehl J.T. ,Grassl H., and Shi L. 1995: Warm pool heat-budget and shortwave cloud forcing — A missing physics? *Science* **267,** 499-503.

Renno, N. , Stone, P.H. , and Emanuel, K. 1994: A radiative-convective model with an explicit hydrological cycle, Part II: Sensitivity to large changes in Solar forcing. *J. Geophys. Res.* (in press).

Sun, D-Z, and Lindzen, R. S. 1993: Water vapor feedback and the ice age snowline record. *Ann. Geophysicae* **11,** 204-215.

Waliser, D. and Graham, N. 1993: Convective cloud systems and warm-pool sea surface temperatures: Coupled interactions and self-regulation. *J. Geophys. Res.* **96,** 15311-15324.

Washington, W. M. and Meehl, G.A 1993: Greenhouse sensitivity experiments with penetrative cumulus convection and tropical cirrus albedo effects. *Climate Dynamics* **8,** 211-223.

Zhang, G. and McPhaden, M. 1995: On the relationship between sea surface termperature and latent heat flux in the Equatorial Pacific. *J. Climate,* in press.

Stratosphere-Troposphere Exchange and its role in the budget of tropospheric ozone

James R. Holton, University of Washington, USA

Jos Lelieveld, Wageningen University, NL

1. Introduction

The stratosphere and troposphere are coupled dynamically, chemically, and radiatively. A particularly important coupling process is the exchange of mass and trace chemical constituents between the stratosphere and the troposphere. The upward transport of tropospheric constituents (both those of natural and those of anthropogenic origin) into the stratosphere initiates much of the chemistry that is responsible for global ozone depletion. The downward transport of stratospheric constituents into the troposphere not only serves as the major sink for some of the constituents involved in stratospheric ozone depletion, but also furnishes reactive species that modify tropospheric chemistry, and provides a source of tropospheric ozone.

Recent advances in our understanding of the dynamical processes controlling the flow of mass into and out of the stratosphere (Haynes et al., 1991) have provided a methodology for estimating the seasonal and hemispheric dependence of stratosphere-troposphere exchange (STE) from conventional stratospheric meteorological observations. Holton (1990) and Rosenlof and Holton (1993) used this methodology to compute seasonal averages for stratosphere-troposphere mass exchange. Their data indicate that the transport of mass out of the stratosphere in midlatitudes is nearly twice as large in the Northern Hemisphere as in the Southern Hemisphere, and is a maximum during Northern Hemisphere winter .

In the classical theory, downward transport from the stratosphere in the extratropics is the dominant source of ozone in the troposphere (Junge, 1962). Crutzen (1973) suggested that the photo-oxidation of carbon monoxide and methane, in the presence of nitrogen oxide, causes large-scale ozone production in the lower atmosphere. Although the question of the relative contributions of downward transport of ozone and in situ photochemical ozone formation in the

NATO ASI Series, Vol. I 35
Clouds, Chemistry and Climate
Edited by Paul J. Crutzen and Veerabhadran Ramanathan

troposphere has still not been completely resolved, several studies suggest that the anthropogenic release of ozone precursors is of major importance in the contemporary troposphere, particularly in the Northern Hemisphere (e. g., Crutzen and Zimmermann ,1991). In this chapter the dynamical basis for recent estimates of STE is first reviewed. A global three-dimensional transport/chemistry model that utilizes these STE rates to examine the relative contributions of STE and tropospheric chemical production to the budget of tropospheric ozone is then described. Finally, results of ozone budget studies with the model are discussed.

2. The dynamics of mean mass exchange

2.1 The dynamical definition of the tropopause

The tropopause is traditionally defined thermally as the lowest level at which the lapse rate of temperature decreases to 2 K km^{-1}, and the average lapse rate from this level to any level within the next 2 km does not exceed 2 K km^{-1}. From a dynamical viewpoint, however, the tropopause is better defined in terms of a specified critical value for a long-lived tracer such as ozone, which has distinctly different stratospheric and tropospheric values. Outside the equatorial zone the dynamical tracer potential vorticity (PV), which is highly correlated with ozone in the extratropical lower stratosphere, is a particularly suitable tracer for defining the tropopause (Hoskins, 1991). PV increases dramatically from troposphere to stratosphere, and it can be readily calculated from conventional wind and temperature data. The PV view of the tropopause can be used to clearly illustrate the fact that the tropopause does not coincide with an isentropic surface, but rather cuts across the isentropes as it slopes downward toward the poles in midlatitudes (see Fig. 1). The region of the stratosphere where the isentropes intersect the tropopause will here be referred to as the "lowermost stratosphere". The lowermost stratosphere must be clearly distinguished from the region above where the isentropes lie entirely in the stratosphere. Following Hoskins (1991) we will refer to this latter region as the "overworld".

2.2 The Brewer-Dobson circulation

In the lowermost stratosphere STE can occur both along and across isentropic surfaces. The mean meridional cross-isentropic mass circulation in the stratosphere is traditionally referred to as the Brewer-Dobson circulation. It is often called the "diabatic circulation" since it is associated with radiative heating

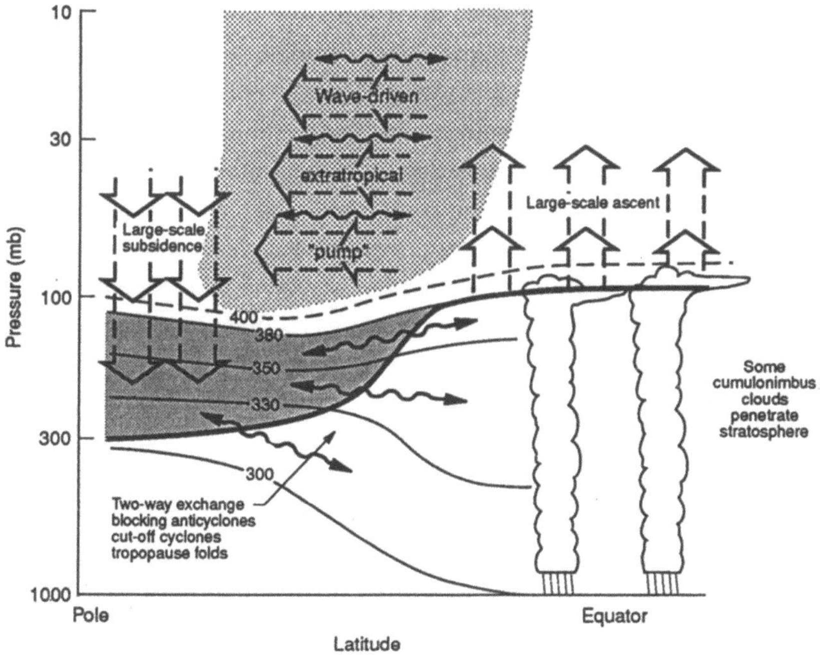

Figure 1: Dynamical aspects of stratosphere-troposphere exchange. The tropopause, defined by constant PV in the extratropics and the 380 K potential temperature surface in the tropics, is shown by the thick line. Thin lines are potential temperature surfaces labeled in Kelvins. The shaded region is the "lowermost stratosphere" where potential temperature surfaces intersect the tropopause, and isentropic exchange by tropopause folding occurs. Light shading in the stratosphere denotes the wave-induced westward zonal force. Wavy arrows indicate quasi-isentropic transport and mixing by large-scale waves. The broad horizontal arrows are the meridional drift that balances the wave-induced zonal force, and the broad vertical arrows show the nonlocally driven equatorial upwelling and extratropical downwelling. (After Holton et al., 1995).

and upward motion across isentropes in the tropics and radiative cooling and downward motion across the isentropes in the extratropics. Recent trace constituent observations by the UARS satellite (Tuck, et al., 1993) confirm earlier evidence from observations of the stratospheric distributions of ozone and water vapor that upward transport in the stratosphere is indeed limited to the tropics, while downward transport occurs in the extratropics, particularly in the winter season. This tropical-up and extratropical-down pattern of meridional overturning tends to produce tracer mixing ratio isopleths that slope downward toward the poles, while mixing along the isentropes by planetary waves tends to flatten the slopes of tracer mixing ratio isopleths.

2.3 Non-local dynamical control of the mass circulation

Although it is associated with diabatic heating and cooling, the Brewer-Dobson circulation is not *forced* by radiative heating, nor is it forced directly from below by penetration of the Hadley circulation into the stratosphere. Rather, the upwelling into the overworld in the tropics is a nonlocal dynamical response to the wave-induced zonal force produced by Rossby waves and gravity waves in the extratropical stratosphere (Haynes, et al., 1991) . This wave-induced force causes a poleward drift of mass across angular momentum surfaces, which is compensated by upward motion accompanied by adiabatic cooling in the tropics and downward motion accompanied by adiabatic warming in the extratropics (see Figure 1). This distribution of adiabatic cooling and heating maintains the temperature below radiative equilibrium in the tropical upwelling region, and above radiative equilibrium in the extratropics. Thus, the distribution of radiative heating and cooling in the stratosphere does not drive the mean meridional mass flow, rather it is a response to the dynamically driven mass flow.

2.4 The annual cycle in STE and its relation to lower stratospheric temperature

The nonlocal control theory of Haynes, et al. (1991) can be applied to the analysis of conventional global meteorological data to produce estimates of the mass flux across a convenient control surface, such as the 100-mb isobaric surface (~ 380 K potential temperature surface in the tropics), which can be regarded as approximating the lower boundary of the overworld. This technique works best for the solstice seasons, when the time change of zonal momentum is small compared to the wave-induced force and the Coriolis force of the mean meridional drift. The nonlocal control method was used with independent data sets to deduce the downward mass fluxes across the 100-mb surface for the Northern and Southern Hemispheres (and by continuity the upward flux in the tropics) by Holton (1990) and Rosenlof and Holton (1993). Results of these two calculations are shown in Table 1. The deduced upward mass flux in the tropics is sufficient to completely replace the mass above the 100 mb surface in about 2–2.5 years. The observations suggest that the mass transport across the tropical tropopause is twice as large in Northern Hemisphere winter as in Southern Hemisphere winter.

TABLE 1: Solstice season mass flux across the 100-mb surface. Numbers labeled (H) from Holton (1990) and numbers labeled (R) from Rosenlof and Holton (1993). Units are 10^8 kg s^{-1}. Negative sign means downward flux. DJF = December, January, February; JJA = June, July, August.

	DJF	JJA	Annual Mean
NH extratropics	(R) -81	(R) -26	(R) -53
	(H) -79	(H) -20	(H) -50
Tropics	(R) 114	(R) 56	(R) 85
	(H) 93	(H) 47	(H) 70
SH extratropics	(R) -33	(R) -30	(R) -32
	(H) -14	(H) -27	(H) -21

This deduced annual cycle in mass transport across the 100-mb surface is consistent with the observed annual temperature cycle near the tropical tropopause. Yulaeva, et al. (1994) showed that this cycle is characterized by temperatures that are several degrees colder in January than in July throughout the tropics, and several degrees warmer in January than in July in the extratropics, as would be expected from the influence of the annual cycle in adiabatic cooling rates. The annual cycle in mass transport is also consistent with the observed cycle in tropical total ozone, which is a minimum in February and a maximum in August, as would be expected from the enhanced vertical advection of ozone poor air into the tropical lower stratosphere during the Northern Hemisphere winter.

2.5 Isentropic exchange in the extratropics

As mentioned above, in midlatitudes the tropopause cuts across isentropic surfaces so that STE can occur through isentropic transport and mixing processes. The boundary between stratospheric and tropospheric air remains very distinct in this region, however; this suggests that there must be a rather strong dynamical resistance to exchange along the isentropes. This resistance is supplied by the mechanism of Rossby-wave propagation. Because of the strong isentropic gradient of PV that marks the tropopause in the lowermost stratosphere, there is a very strong Rossby-wave restoring force in that region, which limits the extent of parcel displacements across the PV gradient. Hence, wave breaking only occurs for large amplitude disturbances. Because there is a large store of available potential energy associated with the strong meridional temperature gradient in this region, large amplitude weather disturbances quite frequently develop in this region, especially in wintertime. The vertical circulations associated with such disturbances create deep intrusions of stratospheric air into the troposphere, which may then become mixed with tropospheric air to produce irreversible transport into the troposphere. Much of the ozone transport from the lowermost stratosphere into the troposphere occurs in connection with such "tropopause fold" events.

The quasi-isentropic exchange initiated by tropopause folding could in theory occur in the absence of the large-scale diabatic circulation. But in that case there would necessarily be an equal quasi-isentropic reverse transport of air from the troposphere into the stratosphere in order to maintain mass balance. The extremely low water vapor mixing ratios observed throughout the stratosphere indicate, however, that the quasi-isentropic exchange in midlatitudes is mostly a one-way transport into the troposphere. Furthermore, the large-scale diabatic circulation is required to transport stratospheric constituents such as ozone downward from the overworld to the lowermost stratosphere. The average rate that such a species can be transported into the troposphere is thus ultimately determined by the rate at which the dynamically controlled large-scale circulation transports mass into the lowermost stratosphere. For this reason the details of mesoscale tropopause fold events may not be important for determining the global flux of ozone from the stratosphere, although they will certainly strongly influence the time and space distribution of such transport.

2.6 The role of tropical convection in STE

Although convection does not control the rate at which the diabatic circulation moves mass into or out of the overworld, penetrative convection acts to "hyperventilate" a layer of a few kilometers depth just above the tropical tropopause. This "tropopause layer" plays an essential role in establishing the stratospheric water vapor budget, and is also important for the irreversible transfer of short-lived constituents across the tropical tropopause. Aircraft and satellite observations of the water vapor distribution in the tropical lower stratosphere reveal the existence of a water vapor mixing ratio minimum (referred to as the "hygropause") that is often well above the tropopause. This feature is now believed to result from dehydration to the saturation mixing ratio at the tropopause, followed by vertical advection into the overworld by the diabatic circulation (Mote et al., 1995). Since, as pointed out earlier, the tropical tropopause temperature is colder in Northern winter than in Northern summer, the driest air enters the stratosphere in Northern winter (when the hygropause is observed to be just above the tropopause), and is advected upward to produce the observed elevated hygropause near 19 km in Northern summer.

3. STE and the budget of tropospheric ozone

3.1. Photochemical ozone formation

In the stratosphere, photodissociation of molecular oxygen ($\lambda < 240$ nm) yields ground state O-atoms that react with O_2 to form ozone. Subsequently, O_3 effectively absorbs solar ultraviolet radiation ($\lambda < 315$ nm), protecting the biosphere from its harmful influence. Nevertheless, a small fraction of the solar UV flux penetrates into the lower atmosphere where it can be absorbed by tropospheric O_3, dissociating it into O_2 and an electronically excited oxygen atom, $O(^1D)$. Most of the $O(^1D)$ atoms dissipate their energy by collisions with air molecules (in troposphere and stratosphere), and the remaining ground state O-atoms recombine with oxygen into O_3. In the humid troposphere, however, $O(^1D)$ atoms have a relatively large chance to react with water vapor, yielding hydroxyl radicals

(1) $O_3 + UV\ (\lambda < 315\ nm) \rightarrow O(^1D) + O_2$

(2) $O(^1D) + H_2O \rightarrow 2OH$

Because these radicals are highly reactive they limit the lifetimes in the atmosphere of many trace gases that are released either by natural or anthropogenic sources (Levy, 1971). Hence, OH is sometimes referred to as the "detergent" of the atmosphere. On a global scale, about one quarter of the OH radicals react with methane (CH_4) and about three quarters with carbon monoxide (CO), i.e.

(3) $CO + HO \rightarrow CO_2 + H$

(4) $H + O_2 + M \rightarrow HO_2 + M$

The collision with M, signifying an air molecule, dissipates excess energy. The hydroperoxy radical (HO_2) can react with O_3

(5) $HO_2 + O_3 \rightarrow OH + 2O_2$

which produces OH, although the ultimate source of OH, ozone (through reactions 1 and 2) is destroyed. In the presence of nitrogen oxide (NO), however, at volume mixing ratios exceeding approximately 10 pptv NO, an alternative reaction path becomes dominant (1 pptv = 10^{-12})

(6) $HO_2 + NO \rightarrow OH + NO_2$

(7) $NO_2 + \text{light } (\lambda < 420 \text{ nm}) \rightarrow NO + O$

(8) $O + O_2 + M \rightarrow O_3 + M$

(Crutzen, 1973). Reactions 6 and 7 regenerate both OH and NO, so that the action of NO is catalytical until it is removed from the atmosphere by chemical destruction or dry deposition ($NO_x = NO + NO_2$).

The global budgets of CH_4, CO and NO_x are strongly affected by anthropogenic sources (Table 2). The emissions of these gases have doubled to tripled during this century, mostly as a result of fossil fuel combustion in northern mid-latitudes. A second important anthropogenic source of trace gases that contributes to in situ O_3 formation is biomass burning in the tropics and subtropics (Crutzen et al., 1979). The maximum burning emissions occur as a result of savanna and forest clearing practices during the months August to October, the dry season in Central Africa and South America. Several measurement campaigns in these areas have shown that O_3 formation in biomass burning plumes can cause regional scale enhancement of O_3 mixing ratios in the boundary layer and also in the free troposphere by up to a factor of 2 (Kirchoff, 1988; Andreae et al., 1992).

TABLE 2 Estimated global emissions of trace gases (Lelieveld and Crutzen, 1994)

Source	CO (Tg/yr)	CH_4 (Tg/yr)	NO_x (Tg N/yr)
Fossil fuel use	450	115	25
Biomass burning	400	30	9
Natural	470	135	11.5
Hydrocarbon oxidation	230		
Ruminants		105	
Rice paddies		70	
Landfills		40	
Sewage		25	
Total	1550	520	45.5

Satellite observations indicate that these plumes sometimes extend over several thousands of kilometers in the Southern Hemisphere (Fishman et al., 1991). In the Northern Hemisphere anthropogenic emissions cause large-scale O_3 build-up during both winter and summer, with particularly strong O_3 formation in the summer. The resulting marked seasonal cycle in surface O_3 concentrations is partly caused by rapid photochemical O_3 production in episodic smog events (Logan, 1985; Feister and Warmbt, 1987; Guicherit, 1988). Re-analysis of 19th century O_3 measurements in comparison with contemporary observations corroborate that anthropogenic emissions have caused a two to three fold increase of surface O_3 in the background continental lower troposphere, and that in the absence of these emissions seasonal variations are only very small (Volz and Kley, 1988). Ozone soundings from the mountain site Payerne, Switzerland, during the past 2 decades, show that also in the free troposphere O_3 has increased significantly, by up to 10-15% per decade in central Europe (Staehelin and Schmid, 1991).

3.2. Transport of ozone

Pre-industrial O_3 concentrations in the northern mid-latitude lower atmosphere maximized during spring (Volz and Kley, 1988). Present-day surface O_3

concentrations in rural locations in the US and Europe, however, are relatively high from April to August, peaking during June/July (Logan, 1985; Guicherit, 1988). The summer maximum is caused by photochemical O_3 formation. The spring enhancement is associated with a buildup of O_3 in the lowermost stratosphere caused by the enhanced downward flux from the overworld during northern winter. This leads to a maximum ozone concentration in the lowermost stratosphere in spring, which in turn leads to a springtime maximum in isentropic transport of O_3 from the lowermost stratosphere into the troposphere by synoptic disturbances.

The quantification of the roles of synoptic-scale and mesoscale STE in the tropospheric O_3 budget, considering that only a relatively small data set from measurement campaigns is available, is still uncertain (WMO, 1986; Vaughan, 1988). The downward mixing of O_3 occurs discontinuously, often on small scales that cannot be resolved by global models. The further transport of O_3 to the lower troposphere is likewise the result of irregular events. Typically, mesoscale subsidence associated with convective storms carries ozone-rich air from the upper troposphere to the surface where it is removed relatively efficiently by chemical destruction, notably through reactions 1 and 2, and by dry deposition on land surfaces (Stallard et al., 1975; Lelieveld and Crutzen, 1994). Also, organized downward flow in extratropical cyclones, e.g. behind cold fronts, contributes to vertical redistribution of ozone.

The degree to which downward transport of O_3 contributes to ozone concentrations, as measured near the surface, has been subject to different interpretation. The pre-industrial spring maximum reported by Volz and Kley (1988) was only weak, whereas recent measurements in not directly polluted background air, e.g. in Finland and in Hawaii, show a pronounced O_3 maximum during the months April and May, which suggests a significant anthropogenic component (Liu et al., 1987; Hakola et al., 1991). Moreover, the abundance of PAN (peroxyacetyl nitrate) in the background troposphere, which is formed in photochemical air pollution, exhibits a concurrent spring peak. It may be that high latitude accumulation of pollutants during the boreal winter, and subsequent southward transport in spring, enhance the natural spring maximum (Penkett and Brice, 1986). On the other hand, in Japan, episodic high O_3 concentrations near the surface have regularly been observed in association with cold fronts; orographically induced (foehn) winds may enhance downward transport of O_3-rich air (Kato et al., 1990). Furthermore, measurements on Barbados and the Canary Islands show a strong anticorrelation between O_3 and anthropogenic pollutants, and a strong positive correlation between O_3 and the stratospheric tracer ^7Be (Savoie et al., 1992; Prospero et al., 1994). In view of these discrepancies we have applied a global transport/chemistry model to estimate the relative importance of in situ photochemistry and STE in the O_3 budget of the global troposphere. It should be emphasized, however, that the description of STE is highly parameterized, reflecting current knowledge about the large-scale features of the atmosphere rather than resolving the above discrepancies.

3.3. Model description

The three-dimensional global model applied has a coarse horizontal resolution of 10 by 10 degrees, and it divides the troposphere into 100-hPa layers up to the 100-hPa level (Zimmermann et al., 1987,1989). Tracer transport is simulated on the basis of monthly mean winds derived from observations (Oort, 1983). The large-scale model diffusion coefficients are proportional to the day-to-day standard deviations of the monthly wind fields. Tracer transport studies with CFC-11, CFC-12 and ^{85}Kr have shown that the large-scale meteorology is well represented by the model (Zimmermann et al., 1989). Rapid vertical transport by deep convection, important for short-lived tracer distributions, has been parameterized with a mass flux scheme (Feichter and Crutzen, 1990). Downward transport of O_3 from the stratosphere is simulated on the basis of pre-calculated air mass fluxes (Holton, 1990) and observed O_3 mixing ratios at the 100 hPa level (Komhyr et al., 1989). The calculated stratospheric outflow of O_3 at the 100-hPa level reaches a maximum in the Northern Hemisphere winter, being about twice as large compared to the summer. Upward transport of O_3 through the tropical tropopause is based on prescribed air mass fluxes (Holton, 1990) and model calculated O_3 in the upper troposphere. The model includes a relatively comprehensive photochemistry routine that accounts for ozone production and destruction pathways in the background troposphere (Crutzen and Zimmermann, 1991). Heterogeneous processes on aerosols and clouds, which reduce the global abundances of tropospheric NO_x and O_3 (Dentener and Crutzen, 1993; Lelieveld and Crutzen, 1990), have been described by Dentener (1993). Model emissions of NO_x (Crutzen and Zimmermann, 1991), CH_4 and CO (Lelieveld and Van Dorland, 1995) include natural and anthropogenic sources; note that the latter exceed the former by roughly a factor of 3 (Table 2). The trace species CO, O_3, NO_2 and nitrate are subject to removal by dry deposition, whereas soluble reaction products are also removed by wet deposition, as described by Crutzen and Zimmermann (1991).

3.4. Results and discussion

The model calculated tropospheric O_3 budget terms are presented in Table 8.3 (Lelieveld and Crutzen, 1994). We have divided the globe into 4 latitudinal zones of equal area, which also approximately reflect the regions directly affected by the Hadley circulation, and the regions north and southward defined as the extratropics. Net O_3 production is subdivided into chemical O_3 formation and loss; the former is determined mostly by reactions of peroxy radicals with NO (e.g. reaction 6); O_3 destruction is dominated by O_3 photodissociation (reactions 1 and 2) and reaction with HO_2 (5), supplemented by O_3 breakdown by OH, and by O_2^- in clouds (Lelieveld and Crutzen, 1990). We calculate that, in total, STE contributes 528 Tg/yr to the global O_3 budget of the troposphere (4.4×10^{10} molec cm^{-2} s^{-1}). The downward transport of O_3 in the extratropics of the Northern Hemisphere exceeds that in the Southern Hemisphere by about a factor of 2; 405 and 186 Tg/yr, respectively (6.7×10^{10} and 3.1×10^{10} molec cm^{-2} s^{-1}). The calculated upward transport of O_3 through the tropical tropopause is 63 Tg/yr, of which only about 15% occurs in the Southern Hemisphere. This is related to the location of the ITCZ, being predominantly north of the equator. Photochemical formation of ozone in the troposphere exceeds transport from the stratosphere by about a factor of 6-6.5. However, photochemical O_3 destruction is also strong, so that the net photochemical O_3 production is 427 Tg/yr, which is about 20% less than net downward transport from the stratosphere. The sum of the net contributions of transport and photochemistry to O_3 in the troposphere is balanced by dry deposition, mostly on land surfaces (Table 3).

TABLE 3:. Calculated O_3 budget of the troposphere (in Tg and Tg/yr) (Lelieveld and Crutzen, 1994)

Latitude	90-30 S	30-0 S	0-30 N	30-90 N	Global
Column O_3	42	57	74	80	253
Strat --> trop	186	-9	-54	405	528
O_3 formation	262	1032	1243	1072	3609
O_3 loss	188	962	1327	705	3182
Net O_3 formation	74	70	-84	367	427
Dry deposition	135	186	261	371	953

As discussed by Vaughan (1988), most estimates of downward O_3 mixing into the troposphere are within the range 360 to 960 Tg/yr (3×10^{10} to 8×10^{10} molec cm^{-2} s^{-1}), which suggests an uncertainty of about a factor of 2. Danielsen and Mohnen (1977) have used potential vorticity (PV) as a proxy for O_3, in combination with air mass fluxes estimated from transport of the radioactive tracer ^{90}Sr released from nuclear bomb tests. Their estimate for hemispheric STE of O_3 is in the upper part of the above range, 7.8×10^{10} molec cm^{-2} s^{-1}. However, this applies to the Northern Hemisphere, whereas the Southern Hemisphere STE flux is expected to be a factor of 2 smaller (Mahlman et al., 1981; Rosenlof and Holton, 1993). Murphy and Fahey (1994) have used the almost linear relationship between O_3 and N_2O in the lower stratosphere, in combination with their estimates of STE fluxes of N_2O based on estimates of stratospheric loss rates; they calculate a global average downward O_3 transport of 3.5×10^{10} molec cm^{-2} s^{-1}. Studies with general circulation models (GCMs), based on fixed PV/O_3 ratios and using PV as tracer, yield global average STE estimates of $3–5\times10^{10}$ molec cm^{-2} s^{-1}, in the lower part of the above range (Mahlman et al, 1980; Gidel and Shapiro, 1979; Allam and Tuck, 1983). Levy et al. (1985) have transported O_3 explicitly in their GCM, and calculate a global mean of 5×10^{10} molec cm^{-2} s^{-1}, in close agreement with our estimate. We also find the same interhemispheric difference.

It should be emphasized that this apparent consensus may not imply that the STE of O_3 is well understood. Ebel et al. (1991) used a regional model of Europe to simulate an STE episode in a cut-off low; their extrapolation to an annual mean on a global scale, yielding $12–15\times10^{10}$ molec cm^{-2} s^{-1}, is much higher than other estimates, although their meteorological simulations apply to the time of year (April) during which STE of ozone appears to be most intense. The regional model of Ebel et al. (1991) resolves relatively small-scale meteorological features, which act on a GCM sub-grid scale, and which may also not be fully accounted for by our nonlocal control estimate (Holton, 1990), but that may nevertheless be important for the temporal and spatial distribution of STE of ozone. However, it must be recalled that ultimately the mean annual flux of O_3 from the stratosphere depends on the rate at which the large-scale diabatic circulation can bring O_3 from the overworld to the lowermost stratosphere, not on the details of individual cut-off lows and tropopause-folding events.

The combination of STE and tropospheric chemistry modeling enables evaluation of the fate of downward transported stratospheric O_3 in the troposphere. Table 8.3 shows that the calculated net STE flux of O_3 is about a factor of 6-7 smaller than O_3 production and O_3 loss by photochemistry. Thus, photochemical production and destruction are more important terms in the O_3 budget of the troposphere than STE. Hence, it may be expected that stratospheric O_3 contributes very little to O_3 concentrations at the earth's surface. By arbitrarily increasing the STE source of O_3 to the troposphere in our model by factors of 2 and 3, we find only marginal changes in surface O_3 (Dentener, pers. comm.). This agrees with the results of a 2-D global model study by Follows and Austin (1992), from which it was concluded that O_3 of stratospheric origin contributes less than 5% to O_3 in the lower troposphere. However, this would be different in a pristine, e.g. pre-industrial troposphere.

Lelieveld and Van Dorland (1995) have calculated that anthropogenically induced photochemistry has increased the O_3 formation and destruction terms in the O_3 budget of the troposphere by more than a factor of 2 during the past 1.5 century. This also shows that anthropogenic O_3 production exceeds the STE flux of O_3 by a factor of 3. However, one should bear in mind that the net global O_3 production term is much smaller (about 427 Tg/yr in the contemporary troposphere; Table 8.3). Furthermore, climate forcings by enhanced infrared radiation absorption resulting from anthropogenically increasing O_3 are particularly strong if O_3 increases occur in the upper troposphere (Lacis et al., 1990). The O_3 budget of the upper troposphere is particularly sensitive to STE of O_3; our calculations show that, on the average, more than half the amount of O_3 above 7 km altitude originates from the stratosphere. The recent anthropogenic breakdown of O_3 in the lower stratosphere, by up to 10% per decade during winter and up to 5% per decade during spring (Staehelin and Schmid, 1990), may thus also affect the upper troposphere, possibly counteracting anthropogenic O_3 increases in the upper troposphere. This underscores the need of reducing the uncertainties associated with STE of ozone.

4. Concluding remarks

There is still considerable uncertainty concerning details of the distribution and magnitude of the flux of ozone from the stratosphere to the troposphere. We have argued, however, that in an average sense this flux must be determined by the rate at which mass is transported downward from the photochemically active region of the stratosphere (the overworld), and hence that the exchange rate can be estimated using mass exchange rates determined from analysis of the large-scale circulation. When utilized in a three-dimensional model for tropospheric ozone, recent estimates of the large-scale exchange rate appear to produce reasonable values for the stratospheric contribution to the tropospheric ozone budget. From Table 3 the replacement time for tropospheric ozone by transport from the stratosphere is about 6 months while the replacement time by in-situ photochemical sources is less than 1 month. The ozone model results thus provide strong evidence that the primary source of tropospheric ozone is photochemical production in the troposphere, not transport from the stratosphere.

References

Allam, R.J. and A.F. Tuck (1984) Transport of water vapour in a stratosphere-troposphere general circulation model. I: Fluxes. *Quart. J. R. Met. Soc.* 110, 321-356.

Andreae, M.O., E.V. Browell, M. Garstang, G.L. Gregory, R.C. Harriss, G.F. Hill, D.J. Jacob, M.C. Pereira, G.W. Sachse, A.W. Setzer, P.L. Silva Dias, R.W. Talbot, A.L. Torres and S.C. Wofsy (1992) Biomass burning emissions and associated haze layers over Amazonia. *J. Geophys. Res.* 93, 1509-1527.

Crutzen, P.J. (1973) A discussion of the chemistry of some minor constituents in the stratosphere and troposphere. *Pure Appl. Geophys.* 106-108, 1385-1399.

Crutzen, P.J., L.E. Heidt, J.P. Krasnec, W.H. Pollock and W. Seiler (1979) Biomass burning as a source of atmospheric gases CO, H_2, N_2O, NO, CH_3Cl and COS. *Nature* 282, 253-256.

Crutzen, P.J. and P.H. Zimmermann (1991) The changing photochemistry of the troposphere. *Tellus* 43AB, 136-155.

Danielsen, E.F. and V.A. Mohnen (1977) Project duststorm report: ozone transport, in situ measurements and meteorological analyses of tropopause folding. *J. Geophys. Res.* 82, 5867-5877.

Dentener, F.J. and P.J. Crutzen (1993) Reaction of N_2O_5 on tropospheric aerosols: impact on the global distributions of NO_x, O_3 and OH. *J. Geophys. Res.* 98, 7149-7163.

Ebel. A., H. Hass, H.J. Jacobs, M. Laube, M. Memmesheimer, A. Oberreuter, H. Geiss and Y.-H. Kuo (1991) Simulation of ozone intrusion caused by a tropopause fold and cut-off low. *Atmos. Environ.* 25A, 2131-2144.

Feister, U. and W. Warmbt (1987) Long-term measurements of surface ozone in the German Democratic Republic. *J. Atmos. Chem.* 5, 1-21.

Fishman, J., K. Fakhruzzaman, B. Cros and D. Nganga (1991) Identification of widespread pollution in the Southern Hemisphere deduced from satellite analyses. *Science* 252, 1693-1696.

Follows, M.J. and J.F. Austin (1992) A zonal average model of the stratospheric contributions to the tropospheric ozone budget. *J. Geophys. Res.* 97, 18047-18060.

Gidel, L.T. and M.A. Shapiro (1979) The role of clear air turbulence in the generation of potential vorticity in the vicinity of upper tropospheric jet stream frontal systems. *J. Atmos. Sci.* 36, 2125-2138.

Guicherit, R. (1988) Ozone on an urban and regional scale. In: I.S.A. Isaksen, *Ozone in the atmosphere*. D. Reidel Publ. Co., Dordrecht, 49-62.

Hakola, H., S. Joffre, H. Lattila and P. Taalas (1991) Transport, formation and sink processes behind surface ozone variability in North European conditions. *Atmos. Environ*, 25A, 1437-1447.

Haynes, P. H., C. J. Marks, M. E. McIntyre, T. G. Shepherd and K. P. Shine, 1991: On the "downward control" of extratropical diabatic circulations by eddy-induced mean zonal forces. *J. Atmos. Sci.*, 48, 651-678.

Holton, J.R. (1990) On the global exchange of mass between the stratosphere and troposphere. *J. Atmos. Sci.* 47, 392-395.

Holton, J. R., P. H. Haynes, M. E. McIntyre, A. R. Douglass, R. B. Rood and L. Pfister (1995): Stratosphere-troposphere exchange. *Reviews of Geophysics*, (in press).

Hoskins, B. J., 1991: Towards a PV-theta view of the general circulation. *Tellus*, 43AB, 27-35.

Junge, C. E., (1962) Global ozone budget and exchange between stratosphere and troposphere, *Tellus*, 14, 363-377.

Kato, H., S.-I. Fujita and S. Nishinomiya (1989) Mechanism of spring high oxidant episode - A meteorological analysis in and around the Hokuriku district, Japan. *Atmos. Environ.* 24A, 2033-2033.

Kirchoff, V.W.J.H. (1988) Surface ozone measurements in amazonia. *J. Geophys. Res.* 93, 1469-1476.

Komhyr, W. D., S. J. Oltmans, P. R. Franchois, W. F. J. Evans and W. A. Matthews (1989) The latitudinal distribution of ozone to 35 km altitude from ECC ozonesonde observations, 1985-1987. In: D. Bojkov and P. Fabian (eds.) *Ozone in the atmosphere*. A. Deepak Publishing, Hampton, V A, 147-150.

Lacis, A.A., D.J. Wuebbles and J.A. Logan (1990) Radiative forcing of climate by changes in the vertical distribution of ozone. *J. Geophys. Res.* 95, 9971-9982.

Lelieveld, J. and P.J. Crutzen (1990) Influences of cloud photochemical processes on tropospheric ozone. *Nature* 343, 227-233.

Lelieveld, J. and P.J. Crutzen (1994) Role of deep cloud convection in the ozone budget of the troposphere. *Science* 264, 1759-1761.

Lelieveld, J. and R. van Dorland (1995) Ozone chemistry changes in the troposphere and consequent radiative forcing of climate. In: W.-C. Wang and I.S.A. Isaksen, *Atmospheric ozone as a climate gas*. D. Reidel Publ., Dordrecht (in press).

Levy II, H. (1971) Normal atmosphere: large radical and formaldehyde concentrations predicted. *Science* 173, 141-143.

Levy II, H., J.D. Mahlman, W.J. Moxim and S.C. Liu (1985) Tropospheric ozone: the role of transport. *J. Geophys. Res.* 90, 3753-3772.

Liu, S.C., M. Trainer, F.C. Fehsenfeld, D.D. Parrish, E.J. Williams, D.W. Fahey, G. Huebler and P.C. Murphy (1987) Ozone production in the rural troposphere and the implications for regional and global ozone production. *J. Geophys. Res.* 92, 4191-4207.

Logan, J.A. (1985) Tropospheric ozone: seasonal behavior, trends and anthropogenic influence. *J. Geophys. Res.* 90, 10463-10482.

Mahlman, J.D., H. Levy II and W.J. Moxim (1980) Three-dimensional tracer structure and behavior as simulated in two ozone precursor experiments. *J. Atmos. Sci.* 37, 655-685.

Mote, P. W., K. H. Rosenlof, J. R. Holton, R. S. Harwood and J. W. Waters (1995) Seasonal variations of water vapour in the tropical lower stratosphere. *Geophys. Res. Lett.* 22 (in press).

Murphy, D.M. and D.W. Fahey (1994) An estimate of the flux of stratospheric reactive nitrogen and ozone into the troposphere. *J. Geophys. Res.* 99, 5325-5332.

Oort, A.H. (1983) Global atmospheric circulation statistics, 1958-1973. NOAA Professional Paper No. 14, US Government Printing Office. Washington, DC.

Penkett, S.A. and K.A. Brice (1986) The spring maximum in photo-oxidants in the northern hemisphere troposphere. *Nature* 319, 655-657.

Prospero, J.M., R. Schmitt, E. Cuevas, D. Savoie, W. Graustein, K. Turekian, A. Volz-Thomas, S. Oltmans and H. Levy II (1995) Temporal variability of ozone and aerosols in the free troposphere over the eastern North Atlantic. Submitted to *Geophys. Res. Lett.*

Rosenlof, K. H. and J. R. Holton, 1993: Estimates of the stratospheric residual circulation using the downward control principle. *J. Geophys. Res.*, 98, 10,465-10,479.

Savoie D.L., J.M. Prospero, S. Oltmans, W.C. Graustein, K.K. Turekian, J.T. Merrill and H. Levy II (1992) Sources of nitrate and ozone in the marine boundary layer of the tropical North Atlantic. *J. Geophys. Res.* 97, 11575-11589.

Staehelin, J. and W. Schmid (1991) Trend analysis of tropospheric ozone concentrations utilizing the 20-year data set of balloon soundings over Payerne (Switzerland). *Atmos. Environ.* 25A, 1739-1751.

Stallard, R.F., J.M. Edmond and R.E. Newell (1975) Surface ozone in the southeast Atlantic between Dakar and Walvis Bay. *Geophys. Res. Lett.* 7, 289-292.

Tuck, A. F., J. M. Russell III and J. E. Harries, 1993: Stratospheric dryness: antiphased desiccation over Micronesia and Antarctica. *Geophys. Res. Lett.*, 20, 1227-1230.

Vaughan, G. (1988) Stratosphere-troposphere exchange of ozone. In: I.S.A. Isaksen (ed.) *Tropospheric ozone*. D. Reidel Publ. Co. Dordrecht, 125-135.

Volz, A. and D. Kley (1988) Ozone measurements made in the 19th century: an evaluation of Montsouris series. *Nature* 332, 240-242.

WMO (1986) Atmospheric ozone 1985. *WMO report 16*, World Meteorological Organization, Geneva.

Yulaeva, E., J. R. Holton and J. M. Wallace (1994) On the cause of the annual cycle in the tropical lower stratospheric temperature, *J. Atmos. Sci.*, *51*, 169-174,.

Zimmermann, P.H. (1987) MOGUNTIA: A handy global tracer model. In: H. van Dop, *Air pollution modelling and its application*. Kluwer, Dordrecht, 593-608.

Zimmermann, P.H., J. Feichter, H.K. Rath, P.J. Crutzen and W. Weiss (1989) A global three-dimensional source-receptor model investigation using ^{85}Kr. *Atmos. Environ.* 23, 25-35.

ENHANCED SHORTWAVE CLOUD RADIATIVE FORCING DUE TO ANTHROPOGENIC AEROSOLS

S. E. Schwartz
Environmental Chemistry Division
Brookhaven National Laboratory
Upton NY 11973 USA

and

A. Slingo
Hadley Centre for Climate Prediction and Research
Meteorological Office, London Road, Bracknell
Berkshire RG12 2SY UK

It has been suggested, originally by Twomey (SCEP, 1970), that anthropogenic aerosols in the troposphere can influence the microphysical properties of clouds and in turn their reflectivity (albedo), thereby exerting a radiative influence on climate. This chapter presents the theoretical basis for of this so-called indirect forcing and reviews pertinent observational evidence and climate model calculations of its magnitude and geographical distribution. We restrict consideration to liquid-water clouds, in part because these clouds are the principal clouds thought to be influenced by anthropogenic aerosols, and in part because the processes responsible for nucleation of ice clouds are not sufficiently well understood to permit much to be said about any anthropogenic influence. The argument for anthropogenic influence on cloud albedo rests on the premise that aerosol particle number concentrations are substantially increased by industrial emissions. As a consequence, the number concentration of cloud droplets N_{cd}, which is determined by the number concentration of aerosol particles in the pre-cloud air, is also increased. This in turn leads to an enhanced multiple scattering of light within clouds and to an increase in the optical depth and albedo of the cloud. In contrast the physical thickness, liquid water content, and liquid water path of the cloud, which are governed to close approximation by large-scale thermodynamics, are considered to be unchanged or at least not greatly influenced by the increase in cloud droplet concentration. There is evidence that this assumption may not be entirely correct. In particular there is indication that the decrease in drop size may inhibit precipitation development, increasing cloud liquid water content and cloud lifetime, both of which effects would contribute

NATO ASI Series, Vol. I 35
Clouds, Chemistry and Climate
Edited by Paul J. Crutzen and Veerabhadran Ramanathan

further to reflection of solar radiation by clouds; this phenomenon is also considered here.

The present chapter reviews the theory of the indirect forcing phenomenon and estimates of its magnitude, field measurements addressing aspects of this phenomenon, and approaches to describing this phenomenon in climate models. This review is necessarily fairly selective. For more extended accounts see Hobbs (1993), Andreae (1995), and Charlson and Heintzenberg (1995).

THEORETICAL BASIS OF INDIRECT FORCING

Aerosol influences on cloud albedo and shortwave forcing. The theoretical basis for a dependence of cloud albedo on cloud droplet number density, and in turn on the number density of the aerosol particles on which cloud droplets form, is outlined by Twomey (1974, 1977a,b). To relate cloud albedo to optical depth and thus to microphysical properties we employ an analytical expression obtained from the two-stream approximation for the reflectance (albedo) R_{CT} of a nonabsorbing, horizontally homogeneous cloud (e.g., Bohren, 1987):

$$R_{CT} = \frac{\delta_C(1-g)}{2+\delta_C(1-g)} \quad . \tag{1}$$

Here g is the asymmetry parameter for single scattering (mean cosine of the scattering angle), approximately 0.85 for cloud droplets of radius much greater than the wavelength of visible light, and δ_C is the optical depth of the cloud,

$$\delta_C = \int dz \int \pi Q_{ext}(r) r^2 N(r) dr \ . \tag{2a}$$

Here $Q_{ext}(r)$ is the extinction efficiency for a cloud droplet of radius r, and $N(r)$ is the size distribution of the cloud drops; the integrals are taken over drop radius and the physical depth of the cloud. For cloud droplets of radius much greater than the wavelength of visible light $Q_{ext}(r)$ may be approximated as a constant, $Q_{ext} \approx 2$ (twice the geometric cross section; Twomey, 1977a). Hence for a uniform cloud

$$\delta_C \approx 2\pi \overline{r^2} N_{cd} z_C \ , \tag{2b}$$

where z_C is the physical thickness of the cloud layer, N_{cd} is the cloud droplet number concentration, and $\overline{r^2}$ is the mean square drop radius given by

$$\overline{r^2} = \frac{1}{N_{cd}} \int r^2 N(r) dr \ .$$

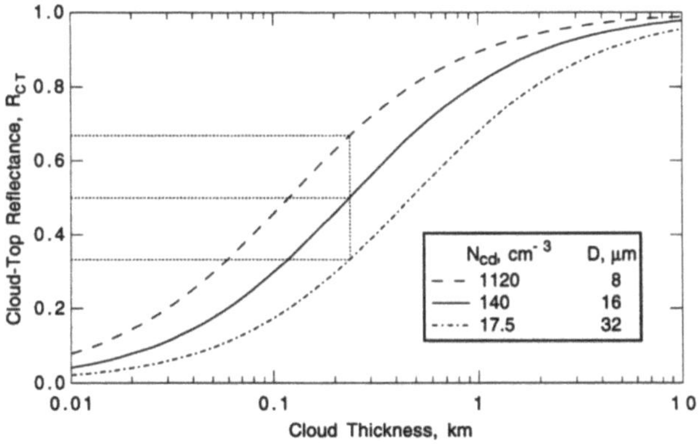

Figure 1. Dependence of cloud-top albedo on cloud thickness for liquid water volume fraction $L = 0.3$ cm^3 m^{-3} and indicated values of cloud droplet number concentration N_{cd}. Note that the sensitivity to an eightfold increase/decrease in N_{cd} (halving/doubling, respectively, of drop diameter) is greatest for clouds of intermediate reflectance $R_{CT} = 0.5$ and falls off for both greater and lesser values of R_{CT}. Modified from Twomey (1977a).

To examine the dependence of R_{CT} on N_{cd} we note that for a cloud of fixed liquid water volume fraction L, the drop radius and number concentration are not independent, but are related by $L = (4\pi/3)\overline{r^3}N_{cd}$, where $\overline{r^3}$ is the mean cube drop radius. Thus, within the approximation $\overline{r^3} = \overline{r^2}^{3/2}$,

$$\delta_C \approx 2\pi z_C \left(\frac{3L}{4\pi}\right)^{2/3} N_{cd}^{1/3} . \tag{3a}$$

More accurately

$$\delta_C = 2\pi z_C \left(\frac{3L}{4\pi}\right)^{2/3} \kappa^{1/3} N_{cd}^{1/3} , \tag{3b}$$

where κ is a dimensionless measure of the dispersion of the drop size distribution

$$\kappa \equiv \overline{r^2}^3 \Big/ \overline{r^3}^2 . \tag{4}$$

For any given shape of the drop size distribution, κ is a constant of value near unity; the value of κ varies slightly for distributions of different shapes.

Equations (1) and (3a) can be used to evaluate the cloud-top albedo as a function of physical thickness of cloud, for specified values of L and N_{cd}, Figure 1, or as a function of N_{cd} for specified values of L and z_C, Figure 2.

Equations (1) and (3a), being analytical, allow an expression to be derived for the sensitivity of cloud-top albedo to a change in N_{cd}, for fixed liquid water content and physical thickness of cloud:

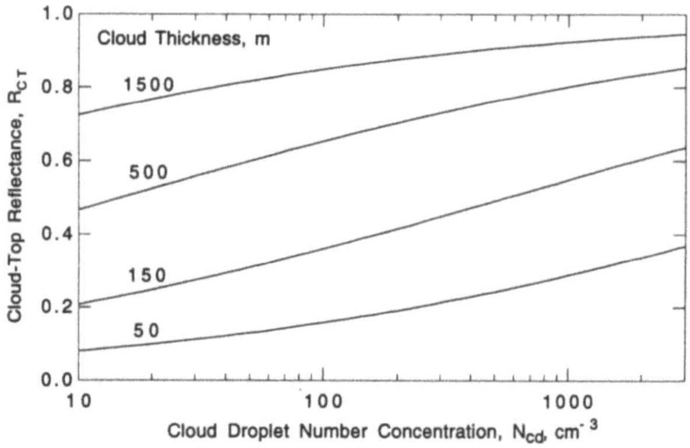

Figure 2. Dependence of cloud-top albedo on cloud droplet number concentration N_{cd} for liquid water volume fraction $L = 0.3$ cm^3 m^{-3} and indicated values of cloud thickness.

$$\Delta R_{CT}\big|_{L,z_C} = \frac{1}{3}[R_{CT}(1 - R_{CT})]\Delta \ln N_{cd} \ . \tag{5a}$$

The quantity $R_{CT}(1 - R_{CT})$ exhibits maximum value of $\frac{1}{4}$ when cloud-top albedo $R_{CT} = \frac{1}{2}$, for which ΔR_{CT} assumes a maximum value $\Delta R_{CT}^{max} = \frac{1}{12}\Delta \ln N_{cd}$. As seen in Figure 3, the sensitivity of cloud-top albedo to change in cloud droplet concentration varies only slowly with R_{CT} for intermediate values of R_{CT}. The approximation

$$\Delta R_{CT}\big|_{L,z_C} \approx 0.075 \Delta \ln N_{cd} \tag{5b}$$

is accurate within 10% for the range of cloud-top albedo $0.28 \lesssim R_{CT} \lesssim 0.72$. This range of albedo is characteristic of the prevalent and climatically important marine stratus clouds.

It should be noted that the above calculations are referred to a fractional change in N_{cd}, $\Delta \ln N_{cd} = \Delta N_{cd}/N_{cd}$. Twomey (1991) introduces what he refers to as an absolute cloud susceptibility,

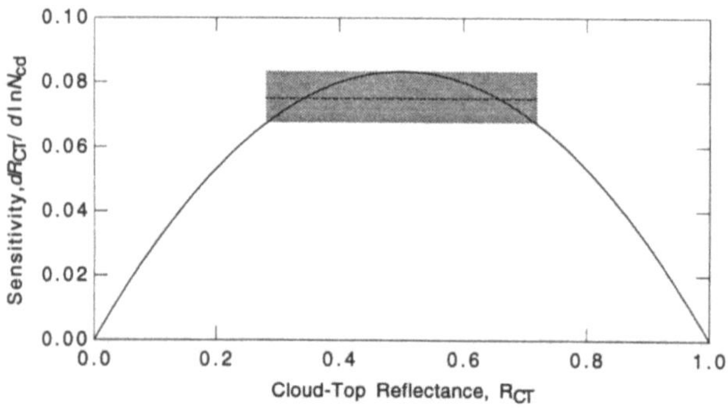

Figure 3. Dependence of sensitivity of cloud-top albedo R_{CT} to a logarithmic change in concentration of cloud droplet number concentration N_{cd}, as a function of R_{CT}. The shaded band indicates the region, $0.28 \lesssim R_{CT} \lesssim 0.72$, for which the approximation $dR_{CT}/d\ln N_{cd} \approx 0.075$ holds within 10%. Modified from Charlson et al. (1992).

$$\frac{dR_{CT}}{dN_{cd}} = \frac{1}{3N_{cd}}\left[R_{CT}(1-R_{CT})\right], \tag{6}$$

and points out that this quantity is greatest in regions where values of N_{cd} are low, such as remote maritime locations. Platnick and Twomey (1994) and Taylor and McHaffie (1994) have evaluated cloud susceptibilities according to this definition, finding not surprisingly, that the susceptibility decreases strongly as the total cloud droplet concentration increases.

Twomey (1977b) carried out numerical calculations to examine the influence that absorbing material in aerosol particles would exert on the albedo enhancement due to increasing droplet concentrations. He found that except for clouds having very high albedo ($R_{CT} \lesssim 0.8$), the albedo enhancement dominated over the albedo decrease due to absorbing material for reasonable assumptions of the magnitude of the absorption.

To gain a sense of the magnitude of albedo change and forcing that would be associated with a possible anthropogenic perturbation in N_{cd}, we note that a 10% relative increase in N_{cd} ($\Delta \ln N_{cd} \approx 0.1$) corresponds to an increase in absolute cloud-top albedo by nearly 1% (0.75%). This high sensitivity to N_{cd}, together with the large amount of shortwave power reflected by clouds, is the basis of the sensitivity of the so-called indirect aerosol forcing effect and may be used to make a initial estimate of the magnitude of global mean radiative forcing that might be attributable to anthropogenic aerosols. Data from the Earth Radiation Budget Experiment (ERBE; Hartmann, 1993) show that the high albedo of clouds, relative to that of the cloud-free surface, increases the global-average reflection of shortwave radiation by about 50 W m^{-2}. If anthropogenic aerosols increase Northern Hemisphere-average cloud droplet number concentrations by 30%, as estimated (Schwartz, 1988) on the basis of comparisons of non-seasalt sulfate concentrations in remote Northern Hemisphere and Southern Hemisphere marine locations, then the resultant cloud brightening due to these aerosols would result in a hemispheric-mean forcing of -1.1 W m^{-2}, although, to be sure, not all such clouds are low-level clouds susceptible to anthropogenic aerosol influence. To place this forcing in the context of greenhouse gas forcing, we note that the longwave radiative forcing by increases in CO_2 over the industrial period is about 1.5 W m^{-2}, and the total for all greenhouse gases over this period is about 2.5 W m^{-2} (Shine et al., 1990). This forcing is also comparable to recent estimates of the direct shortwave forcing due to light scattering by anthropogenic aerosols in cloud-free conditions (Charlson et al., 1991, 1992; Kiehl and Briegleb, 1993; Taylor and Penner, 1994). Thus the indirect forcing by anthropogenic aerosols, as well as the direct forcing, needs to be taken into account in the context of anthropogenic influences on climate (Wigley, 1994; IPCC, 1995).

A somewhat different approach to estimating the global mean forcing due to anthropogenic aerosols was taken by Charlson et al. (1992; cf. also Kaufman et al., 1991), who attempted to assess the global-mean forcing based on the fractional coverage of marine stratus clouds. For a given cloud-top albedo perturbation the corresponding perturbation in top-of-atmosphere (TOA) albedo is

$$\Delta R_{\mathrm{TOA}} = T^2 \Delta R_{\mathrm{CT}} \tag{7}$$

where T is the fraction of incident shortwave radiation transmitted by the atmosphere above the cloud layer. To obtain the change in global- or hemispheric-mean albedo due only to the change in albedo of marine stratus clouds $\Delta R_{\mathrm{TOA}}^{\mathrm{mst}}$, the albedo change given by (7) must be decreased further by the fractional coverage by marine stratus clouds, A_{mst}, yielding

$$\overline{\Delta R_{\mathrm{TOA}}} = A_{\mathrm{mst}} \Delta R_{\mathrm{TOA}}^{\mathrm{mst}} . \tag{8}$$

The corresponding perturbation in global- or hemispheric-mean shortwave forcing was estimated as

$$\overline{\Delta F_{\mathrm{C}}} = -\overline{F}\,\overline{\Delta R_{\mathrm{TOA}}} = -(F_{\mathrm{T}}/4)\overline{\Delta R_{\mathrm{TOA}}} , \tag{9}$$

where $\overline{F}$ is the global- or hemispheric-mean top-of-atmosphere shortwave radiation evaluated as $\overline{F} = F_{\mathrm{T}}/4$, where F_{T} is the solar constant. The negative sign denotes a cooling tendency. The forcing depends linearly on the perturbation in cloud-top albedo and thus exhibits a logarithmic dependence on N_{cd} given by (5).

$$\overline{\Delta F_{\mathrm{C}}} \approx -0.075(F_{\mathrm{T}}/4) A_{\mathrm{mst}} T^2 \Delta \ln N . \tag{10}$$

Figure 4 shows the dependence of global- or hemispheric-mean radiative forcing on a change in N_{cd} and indicates a sensitivity comparable to that estimated above. For an assumed 30% increase in hemispheric-mean N_{cd}, the hemispheric-mean forcing evaluated by this approach is -1.1 W m^{-2}, identical to that obtained using the cloud radiative forcing determined from the ERBE data.

As discussed below, more detailed model calculations lead to estimates of similar magnitude, although it must be emphasized that the premise of the estimate

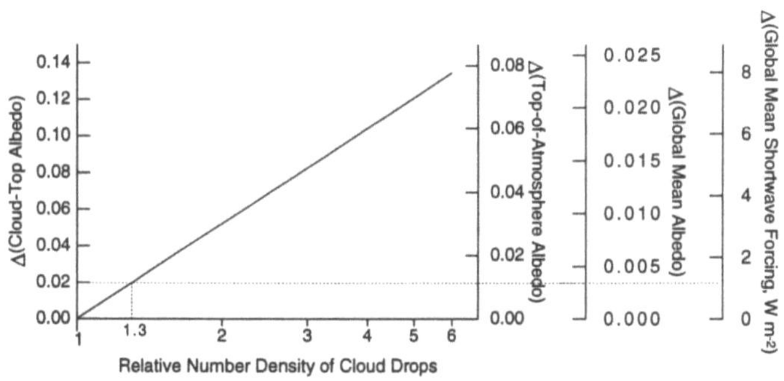

Figure 4. Calculated perturbation in cloud-top albedo (left ordinate), top-of-atmosphere albedo above marine stratus, global-mean albedo, and global-mean cloud radiative forcing (right ordinates) resulting from a uniform increase in cloud droplet number concentration N_{cd} by the factor indicated in the abscissa. The global-mean calculations were made with the assumption (Charlson et al., 1987) that the perturbation affects only non-overlapped marine stratus and stratocumulus clouds having a fractional area of 30%; the fractional atmospheric transmittance of shortwave radiation above the cloud layer was taken as 76%. The dotted line indicates the perturbations resulting from a 30% increase in N_{cd}. Modified from Charlson et al. (1992).

(30% hemispheric enhancement in cloud droplet number concentrations) remains little more than an educated guess at present. Much work needs to be done before the uncertainty associated with the indirect forcing by anthropogenic aerosols, as well as that associated with the direct forcing, can be reduced to a value comparable with that associated with anthropogenic greenhouse gases (Penner et al., 1994).

Theoretical basis for anthropogenic influence on cloud droplet concentrations. Having set forth the basis for the dependence of R_{CT} on N_{cd}, we turn to examination of the influence of anthropogenic aerosols on cloud droplet concentrations N_{cd}. In this examination we rely mainly on observations, but with guidance from theory (e.g., Pruppacher and Klett, 1980; Hänel, 1987; Hobbs, 1993) and model studies (Jensen and Charlson, 1984; Flossmann et al., 1985; Ahr et al., 1989).

First we give a quick overview of the basis for this influence. Briefly clouds form when an air parcel, which inevitably contains aerosol particles as well as water vapor, is cooled to a temperature below its dew point and water vapor condenses on existing aerosol particles to form cloud droplets. As the temperature decreases to and below the dew point, the ambient water vapor pressure exceeds the equilibrium water vapor pressure characterizing the aqueous (hygroscopic or deliquescent) aerosol particles present in the air. This departure from equilibrium leads to net condensation of water on the particles. The condensation on any given particle tends to be resisted by the energy required to create the additional surface area of the growing droplet and favored by the vapor pressure reduction resulting from the presence of soluble material in the particles. The combined influence of these competing effects results in a free-energy barrier that is characterized by a critical drop radius and vapor pressure whose values depend on the solute mass and composition. The value of the critical vapor pressure decreases with increasing solute mass because of surface-to-volume considerations.

During the initial stage of droplet growth the surface free-energy term is dominant, and the accretion of additional water by a droplet increases its equilibrium water vapor pressure, tending to inhibit further growth. However, if the ambient vapor pressure of water exceeds the critical vapor pressure of a given particle for sufficiently long that the radius of the nascent cloud drop grows beyond its critical value, then further droplet growth decreases its equilibrium water vapor pressure, and consequently the droplet will continue to grow even as the ambient vapor pressure subsequently decreases. This process is referred to as cloud droplet "activation." After the initial peak in supersaturation in the early stage of cloud formation, the surface area of the growing drops provides an increasing sink for water vapor and the ambient supersaturation decreases. Larger particles, which have been activated, continue to grow; smaller particles, which were not activated, shrink and remain as "interstitial" aerosol particles.

The number of aerosol particles activated during cloud formation depends on the number of aerosol particles present in the pre-cloud air that were sufficiently large, and containing a sufficient amount of hygroscopic material, to have become activated given the time history of supersaturation at the early stage in the cloud cycle. This supersaturation time profile depends in the first instance on the time profile of the thermodynamic driving force--the updraft velocity and the resultant rate of generation of supersaturation, or, more precisely, the rate of decrease of the saturation vapor pressure of water. Thus the number concentration of cloud droplets might be expected to increase with the number concentration of aerosol particles of a size range suitable to become activated for a given maximum supersaturation. However this dependence is not a simple one-to-one relation because the supersaturation time profile itself depends on the aerosol loading, since a greater concentration of particles and the associated surface area available for water condensation will lead to a reduced ambient supersaturation for a given thermodynamic generation rate. One will thus expect a decrease from a one-to-one relation as aerosol particle concentrations become sufficiently great. However the onset of the fall off may be expected to depend on updraft velocity, insofar as this variable controls the generation rate of supersaturation. It is in the context of this picture that we examine the influence of perturbations on cloud droplet concentrations due to anthropogenic aerosols.

In consideration of anthropogenic influences on CCN and cloud droplets much attention has been focused on sulfur compounds, since sulfate comprises a major fraction of accumulation-mode (radius ~0.1 to 1 μm) aerosol particles and since sulfuric acid and its neutralization products (collectively, sulfate) are highly water-soluble and therefore effective atmospheric aerosols. The mechanism by which SO_2, the major emitted species, is oxidized to form sulfate and by which the aerosol forms and grows to a size that is effective as CCN at modest supersaturations is not fully established. SO_2 is oxidized in the gas phase and hydrated to form sulfuric acid monomer, which, because of its low vapor pressure, adds to existing particles or forms new particles depending on the formation rate and the available surface area. However concerns have been raised that the aerosol growth process is too slow to account for particles that serve as CCN at low updraft velocities characteristic of marine stratus clouds. SO_2 is oxidized also in aqueous solution, in clear-air particles and in cloud droplets. It is of course tautological that concentrations of aerosol particles can increase only by mechanisms involving new particle formation. Nonetheless aqueous-phase oxidation may contribute substantially to the process by which particles grow to a size sufficiently great to serve as CCN in marine stratus (Fitzgerald, 1991; Hegg, 1990; Lin et al., 1992; Lelieveld and Heintzenberg, 1992; Kaufman and Tanré, 1994). Other anthropogenic materials that are thought to contribute substantially to CCN are nitrates, i.e., nitric acid and its salts, (Kulmala et al., 1993; Wurzler et al., 1995) and organics, including those from biomass combustion (Warner and Twomey, 1967; Novakov and Penner, 1993; Andreae, 1995). On the other hand, there is a growing body of evidence that, depending on composition, not all particles of a given size serve equally well as CCN, and that even in a given cloud the fraction of insoluble material present on particles that are activated to form cloud drops increases with increasing particle size (Hansson and Svenningsson, 1994; Hallberg et al., 1994).

Influence of cloud droplet concentrations on cloud lifetime. An additional influence of anthropogenic CCN on shortwave cloud radiative forcing may come from an increase in cloud lifetime. The formation of precipitation within clouds is strongly dependent on the size distribution of cloud droplets; warm precipitation develops much more efficiently for a given liquid water content in clouds with fewer, larger drops than in clouds with more, smaller drops. Albrecht (1989) described potential increases in cloud lifetime if precipitation rates were reduced by a decrease in cloud droplet size due to anthropogenic CCNs and called attention to the climatic implications of this effect. Ackerman et al. (1993) presented model results describing the collapse of stratus cloud formation processes from a deficit of CCN, providing further indication that additional CCN could stabilize a cloud against dissipation by drizzle. This phenomenon gains support by the occurrence of ship tracks in cloud free regions.

In addition to the shortwave radiative influence, enhanced persistence of clouds due to anthropogenic aerosols can also exert a longwave influence, since a more persistent cloud will absorb and radiate in the thermal infrared as well as reflect shortwave radiation. Further, to the extent that water that would otherwise be precipitated remains in the atmosphere as vapor, that water vapor will exert an additional greenhouse influence. It thus appears that representation of these effects in climate models will require a considerable extension of the treatment of cloud processes beyond that in which cloud processes are governed by large scale thermodynamics alone.

OBSERVATIONAL STUDIES

We review here observations relating mass concentrations of aerosol constituents and number concentrations of aerosol particles, cloud condensation nuclei, and cloud droplets. Unfortunately there are few studies relating all of these variables; typically only two quantities are measured at a time, and thus it is not possible to characterize the linkage among all these quantities even in a single study, and all the more in any climatologically meaningful way. On the other hand there are a large number of studies looking at pieces of the picture; the studies described here should be considered as representative of that larger set.

Anthropogenic influences on CCN concentrations. A key determinant of aerosol influences on cloud droplet concentrations is the number concentration of cloud condensation nuclei, CCN, particles which will activate to form cloud droplets at a given supersaturation, maintained indefinitely. CCN concentrations are thus not equal to cloud droplet number concentrations, but are a convenient means of characterizing the cloud nucleating ability of a given air sample. CCN concentrations may be determined experimentally by exposing an ambient air sample to a known supersaturation (typically 0.3% to 1.5%) for sufficiently long to activate all particles whose critical supersaturation is less than or equal to the imposed supersaturation and counting the resultant droplet concentration, which is manifest by the enhanced light scattering. (Twomey, 1977a; Twomey et al., 1978). In a given air sample the CCN number concentration increases with increasing supersaturation, consistent with the depiction of the activation process given above. Because of this dependence it is necessary to specify the supersaturation characterizing a given CCN concentration measurement and to compare measurements only at a specified supersaturation.

It has been known for some time that CCN concentrations are greater in continental air masses than in marine air masses--by a factor of perhaps three or more. Summarizing numerous studies, Pruppacher and Klett (1980) conclude that CCN concentrations in maritime air uninfluenced by anthropogenic emissions rarely exceed 100 cm^{-3}, whereas in air that has been over land for several days concentrations in excess of 1000 cm^{-3} are frequent. As discussed below, however, it is clear that much of the so-called continental enhancement must be attributable to anthropogenic influences. With respect to natural CCN, Twomey (1977a) reports observations of vertical profiles of CCN concentrations over Colorado (surface concentration 230 cm^{-3} at 0.35% supersaturation) and the Caribbean (80 cm^{-3}). At both locations CCN concentrations decrease with increasing altitude, indicative of a surface source. On the basis of such observations and the lack of any indication of direct emission of CCN from surfaces, Twomey suggests that the principal natural source of CCN is particle formation in the atmosphere from gases emitted at the surface and that these emissions are greater over vegetated land surfaces than over oceans. Mészáros (1992), likewise, considers the surface to be the principal source of CCN, but hypothesizes that in the absence of anthropogenic sources, continental CCN concentrations may be substantially less than marine. For another view, however, see Slinn (1992) and Raes (1995), who argue on behalf of the contribution of natural CCN formed in the free troposphere from gas to particle conversion and subsequent particle growth. Ayers and Gras (1991) present data obtained in air minimally influenced by anthropogenic emissions at Cape Grim, Tasmania, indicative of a dependence of CCN concentrations (50 to 150 cm^{-3} at 0.2% supersaturation) on the concentration of methanesulfonic acid, an oxidation product of marine dimethylsulfide (DMS).

Table 1. Summary of CCN concentration data upwind and downwind of St. Louis (Braham, 1974)

Dates	Sampling level (m, msl)	Number of upwind-downwind comparisons	Number of days downwind CCN (0.5%) > upwind		CCN concentrations at various supersaturations 0.35%	0.5%	1.0%
Jul-Aug, 1971	610	19	19	$N(u)$	1071	1370	2210
				$N(d)$	1770	2313	3890
				$N(d)/N(u)$	1.65	1.69	1.76
Jan-Mar, 1972	460	13	12	$N(u)$	355	477	848
				$N(d)$	606	857	1678
				$N(d)/N(u)$	1.71	1.80	1.98
Jul-Aug, 1972	top of mixed layer to 460	25	23	$N(u)$	602	753	1166
				$N(d)$	821	1036	1625
				$N(d)/N(u)$	1.36	1.38	1.40
Mar-Apr, 1973	460	10	9	$N(u)$	357	502	977
				$N(d)$	467	686	1451
				$N(d)/N(u)$	1.31	1.37	1.49

In any event, in addition to natural CCN it is well established that CCN concentrations are greatly enhanced in the vicinity of industrial activities, especially combustion, that can release massive amounts of aerosol particles and/or gaseous precursors into the atmosphere. For example Twomey (1974; see also Twomey et al., 1978) reported measurements of CCN concentrations at a site in southeastern Australia. Under conditions of air flow from "relatively clean" land CCN concentrations (supersaturation 0.75%) were as low as 50 cm^{-3}, but when the air had passed over an industrial area the concentration increased to 4500 cm^{-3}. In a study of the influence of urban emissions on CCN concentrations Braham (1974) found substantial enhancement of CCN concentrations in comparisons of concentrations upwind and downwind of St. Louis, Missouri, as summarized in Table 1. CCN concentrations were substantially enhanced downwind of the city on almost all days studied, with the average enhancement factor ranging from 1.31 to 1.98. Figure 5 shows a comparison of supersaturation spectra (concentration of CCN versus percent supersaturation) obtained upwind and downwind of the city; at any given supersaturation the average enhancement in concentration was about a factor of 2.

In aircraft measurements off the eastern seaboard of the United States, Radke and Hobbs (1976) reported CCN concentrations (at 0.2% supersaturation) of 1000 to 3500 cm^{-3} under conditions of northwesterly flow such that the air mass sampled was representative of air blowing off the continent. They compared these

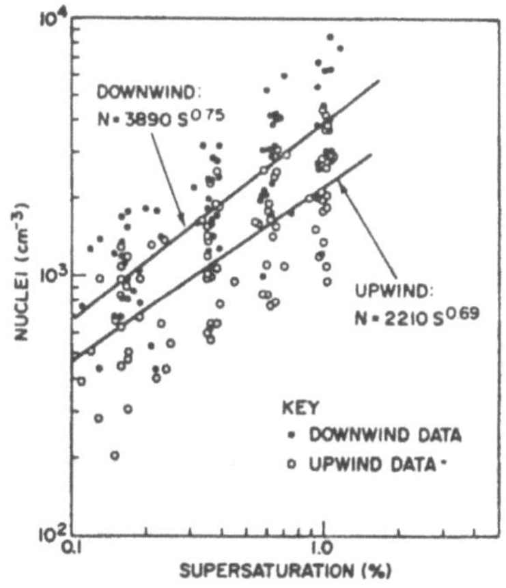

Figure 5. Dependence of CCN concentration on supersaturation measured in 19 upwind-downwind pairs of air samples collected on aircraft flights the vicinity of St. Louis, Missouri during July and August, 1971. The straight lines are log-linear regression fits to the data. (Fitzgerald and Spyers-Duran, 1973).

concentrations to those in clean maritime air masses on the Pacific Coast of Washington state (typically less than 100 cm^{-3}) or over the high plains of the United States (about 300 cm^{-3}). These results were interpreted as indicative of the dominance of anthropogenic CCN in industrial or heavily populated areas, although the investigators explicitly refrained from implying that regions far from large anthropogenic sources of CCN are appreciably affected by CCN from anthropogenic sources, in view of the short (several-day) average residence time of CCN in the atmosphere. More recently Hudson (1991) has reported measurements along the coast of California indicating a background marine concentration (0.7% supersaturation) of 20 to 40 cm^{-3}, compared to "non-urban" concentrations 100 km inland, in Oregon, of 100 to 200 cm^{-3} and to "urban" concentrations in the vicinity of Santa Cruz, California of 3000 to 5000 cm^{-3}. Frisbie and Hudson (1993) report CCN concentrations (0.9% supersaturation) at 800 hPa, upwind and downwind of Denver, Colorado of 500 and 5000 cm^{-3}, respectively. These and other measurements give ample evidence of the substantial anthropogenic enhancement of CCN concentrations. It is clear, with reference to equation (10) or Figure 4, that if cloud droplet concentrations were to exhibit anthropogenic enhancement by factors such as those reported in these measurements of CCN concentrations, the perturbation on cloud albedo and the resultant radiative forcing would be enormous.

The high sensitivity of cloud albedo and radiative forcing to cloud droplet concentration thus points to a key issue, namely, the extent to which CCN and cloud droplet concentrations are enhanced by anthropogenic aerosols over broad regions of the marine atmosphere. This issue remains quite unresolved.

Schwartz (1988) proposed on the basis of measurements in the North Atlantic and rather limited measurements in the North Pacific that concentrations of non-seasalt sulfate were 30% greater in the Northern Hemisphere (NH) marine environment than in the Southern Hemisphere (SH), and by inference that the same held for CCN. This argument was challenged by Charlson et al. (1989) with respect to both the measurement database and the assumption of a linear relation between sulfate loadings and CCN concentrations. However based on long-term measurements of sulfate and methanesulfonate at a network of Pacific islands, Savoie and Prospero (1989) suggested an annual average anthropogenic enhancement of sulfate in the mid North Pacific of about 20%. On the other hand, in a review of aerosol properties in the marine environment, Fitzgerald (1991) presented evidence of a rapid shift from continental to marine aerosols with increasing distance from continents. In any event it is clear that the high sensitivity of radiative forcing to increased cloud droplet concentrations (equation 10; Figure 4) not only makes it crucial that the magnitude of the anthropogenic perturbation be known with some confidence but also places stringent requirements on the ability to determine this quantity.

One final point pertinent to this issue and to the accurate representation of the indirect forcing due to anthropogenic aerosols in climate models concerns the episodic nature of the anthropogenic perturbation in aerosol loading. Because of the concentration of sources in a few highly localized industrial regions, the short residence times of aerosol particles in the troposphere (about a week), the episodic nature of removal processes, which are dominated by precipitation, and the variability of transport winds, concentrations of anthropogenic aerosols at any given location are highly variable temporally. This property is evidenced by time series in a variety of locations of concentrations of aerosol substances and of other indicia of aerosol loadings such as light scattering coefficient and optical depth. In a recent study (Benkovitz et al., 1994) in which the column burden of sulfate aerosol in the mid North Atlantic was calculated with a model that used meteorological variables derived from operational meteorological products, the characteristic autocorrelation time was about 10 hours. The spatial autocorrelation distance was likewise rather short, 400 to 1700 km. This situation is very different from the rather smooth temporal and spatial patterns that are depicted in modeling studies based on monthly average meteorological fields (e.g., Langner and Rodhe, 1991), which have served as the basis for evaluations of the anthropogenic indirect influence, as discussed below. The highly variable nature of the aerosol loading has implications regarding such evaluations in view of the nonlinear relation between aerosol loading and the indirect forcing. It also suggests that it may be easier to discern the anthropogenic influence on aerosol loadings by measures that take account of this episodicity than simply by examination of annual or monthly mean loadings.

Anthropogenic influences on cloud droplet concentrations. Anthropogenic influences on cloud droplet concentrations can be examined directly, by consideration of cloud droplet concentrations in air which has been classified according to anthropogenic influence. Alternatively one may examine the relation between aerosol loading and cloud droplet concentrations. While the former approach can certainly establish the existence of such an influence, the latter approach may ultimately prove more valuable by leading to a quantitative relation between anthropogenic emissions and cloud droplet loadings through mechanistic understanding of the governing processes.

It has been recognized for several decades that continental clouds exhibit greater number concentrations than do marine clouds (e.g., Pruppacher and Klett, 1980). Although the reasons for this are not entirely established, it would appear that to great extent the difference lies in the concentration of CCN in precursor air. In principle cumulus clouds with greater updraft velocities (and therefore greater supersaturation rates) would be expected to exhibit greater droplet number concentrations than stratiform clouds. This expectation is in general borne out, and therefore comparisons must be made for similar cloud types. Still continental clouds tend to exhibit greater drop concentrations. This difference was thought by some to be due to natural sources of aerosols, e.g., particles formed from gas-to-particle conversion of emissions from terrestrial vegetation. However numerous studies linking high cloud drop concentrations to identified combustion sources or other industrial activity have led some to suppose that high cloud drop concentrations in continental clouds may be due largely to anthropogenic emissions (e.g., Mészáros, 1992). We note here a number of these studies.

In a key early study Warner and Twomey (1967) found that the number concentration of droplets in clouds formed downwind of sugar cane fires was greatly increased (average 510 cm^{-3}) over that in maritime air upwind (average 104 cm^{-3}). In an early study examining urban influences on cloud microphysical properties Fitzgerald and Spyers-Duran (1973) compared the droplet size spectra in small cumulus and stratocumulus clouds upwind and downwind of St. Louis, Missouri. Comparisons on two days are shown in Figure 6. It was found that

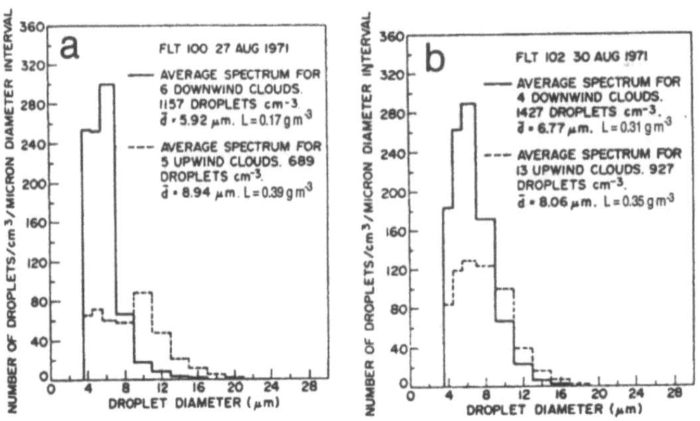

Figure 6. Drop size distributions obtained in cumulus and stratocumulus clouds up- and downwind of St. Louis, Missouri. Note increase in drop concentration and decrease in drop diameter downwind of the city. From Fitzgerald and Spyers-Duran (1973).

clouds downwind of the city were comprised of higher concentrations of smaller droplets that exhibited a narrower dispersion of sizes than upwind clouds. A possible concern with attributing the differences entirely to microphysical changes due to changed CCN concentrations is the difference in liquid water content, at least in the 27 August data.

A rather direct examination of the influence of urban emissions on the microphysical properties on stratiform clouds was recently conducted by Alkezweeny et al. (1993), who reported measurements of cloud droplet size distributions upwind and downwind of the Denver, Colorado metropolitan area. An example of data from this study is shown in Figure 7a. Here the solid curve

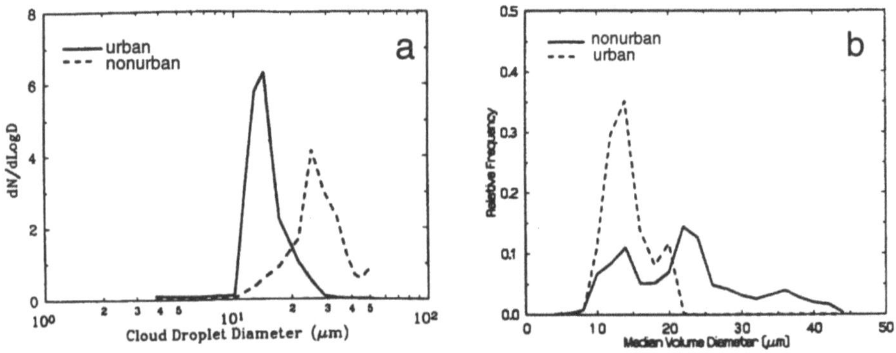

Figure 7. (a) Normalized drop size distributions obtained in vertical profiles in regions of stratiform cloud that were or were not influenced by emissions from Denver, Colorado. (b) Median volume diameter of drop size distributions in stratiform clouds in the vicinity of Denver, according to criterion $NO_2 < 0.5$ ppb (nonurban) or $NO_2 > 1$ ppb (urban). From Alkezweeny et al. (1993).

represents a composite drop-size spectrum obtained in a vertical profile in urban-influenced air and the dashed curve represents similar measurements in nonurban air a few minutes later. The liquid water contents were fairly similar, 0.25 and 0.16 g m^{-3}, respectively. However the drop distribution in the urban air is shifted to much smaller diameters than that of the non urban air, with median volume diameter 14 and 28 μm, respectively. Correspondingly the drop concentration is increased by an order of magnitude, from 22 to 226 cm^{-3}. Figure 7b shows a summary of median volume diameters for the study as a whole; here the criterion for urban versus nonurban air was taken as NO_2 mixing ratio greater than 1 ppb or less than 0.5 ppb, respectively. Despite the somewhat arbitrary nature of the criterion, it is clear that clouds in "urban" air frequently exhibit a considerably reduced drop size relative to clouds in the control "nonurban" air. These measurements likewise establish a strong sensitivity of drop number and drop size to urban emissions.

Evidence of the influence of enhanced aerosol particle concentrations on the concentration of cloud droplets in marine stratus and stratocumulus clouds comes from so-called "ship tracks", linear features of high cloud reflectivity embedded in such clouds that apparently form on aerosol particles released in the exhaust from the engines of ships (Coakley et al., 1987; Scorer, 1987; Radke et al., 1989). Aircraft penetrations through ship tracks have confirmed the supposition of enhanced droplet concentrations and decreased drop radius associated with the ship tracks compared to adjacent unperturbed regions of the clouds: 120 cm^{-3} vs. 40 cm^{-3} and 7.5 μm vs. 11.2 μm, respectively (Radke et al., 1989; King et al., 1993).

In an examination of anthropogenic influences on cloud microphysical properties at somewhat greater distances from sources, Pueschel et al. (1986) reported measurements of microphysical properties and chemical composition of clouds at Whiteface Mountain in rural upstate New York (elevation 1500 m above sea level). Table 2 shows a comparison of measurements on three days for which airmass back trajectories indicated distinctly different sources of the air in which the clouds were formed. The cloud droplet concentration increased dramatically with aerosol loading, which, in turn, on the basis of the back trajectory, was ascribed to anthropogenic emissions, certainly in the case of the trajectory that passed in the vicinity of Pittsburgh, Pennsylvania, compared to the background continental and maritime trajectories.

Table 2. Comparison of accumulation-mode aerosol particle concentration (0.2-1.0 μm diameter), cloud droplet concentration (0.5-47 μm diameter), and sulfate concentration in cloudwater as a function of air-mass back trajectory for samples at Whiteface Mountain, New York, in June 1981. Data of Pueschel et al. (1986).

Date, 1981	Air-mass back trajectory	N_{cd} cm^{-3}	N_{cd}, cm^{-3}	$F = N_{cd} / (N_{cd} + N_{cd})$	Sulfate , μmol L^{-1}
26 June 0819-0836	Maritime; St. Lawrence Estuary, Gaspé Peninsula	63	64	0.50	6.5
23 June 0845-0944	Background continental; north of Lake Superior	219	127	0.37	21
25 June 0859-0955	Polluted continental; Pittsburgh, Pennsylvania	3060	750	0.20	237

Several studies have reported relations between cloud drop concentration and concentrations of aerosol particles or of CCN in pre-cloud air or interstitial cloud air, from which inferences may be drawn regarding the sensitivity of cloud drop concentrations to anthropogenic influences. Figure 8 shows a plot of cloud droplet concentration vs. below-cloud concentration of CCN active at an updraft velocity of 3 m s^{-1} presented by Twomey and Warner (1967). An approximately one-to-one relation was observed over the entire concentration range studied, up to about 1000 cm^{-3}.

Considerably different results were obtained by Pueschel et al. (1986) in their examination (at Whiteface Mountain, New York) of the sensitivity of cloud droplet concentration to increasing total in-cloud particle concentration, N_{tot} =

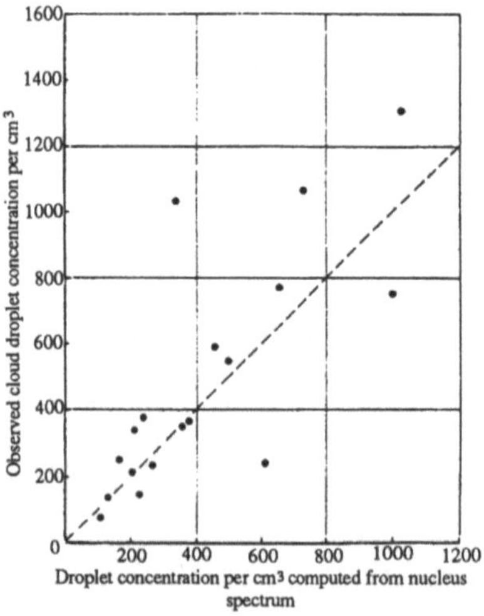

Figure 8. Dependence of cloud droplet concentration N_{cd} on below-cloud concentration of CCN active at vertical velocity of 3 m s^{-1} or less. Dashed line shows one-to-one correspondence. From Twomey and Warner (1967).

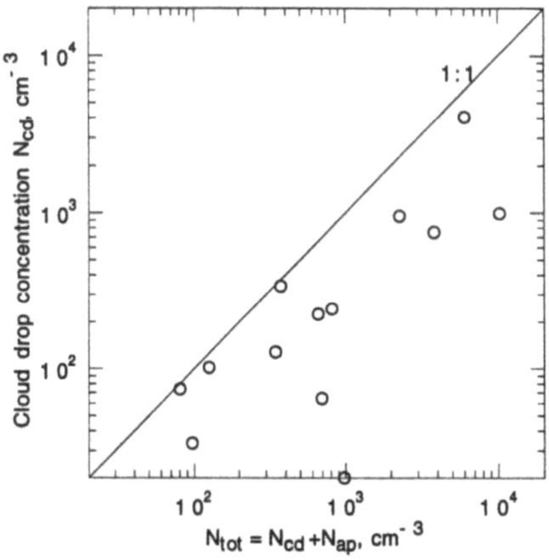

Figure 9. Dependence of cloud droplet number concentration N_{cd}, on total particle number concentration $N_{tot} = N_{cd} + N_{ap}$. Here cloud droplets (cd) encompass the range 0.5 to 47 µm diameter and accumulation-mode aerosol particles (ap) the range 0.1 to 3 µm diameter. Data of Pueschel et al. (1986) obtained at Whiteface Mountain in rural upstate New York (elevation 1500 m above sea level) in June 1981.

$N_{cd} + N_{ap}$; here cloud droplets (cd) denote particles in the diameter range 0.5 to 47 µm and accumulation-mode aerosol particles (ap) denote particles in the diameter range 0.1 to 3 µm. The results are summarized in Figure 9. Quite wide ranges were encountered of both cloud droplet concentrations (up to 4000 cm^{-3}) and total particle concentrations (up to 10^4 cm^{-3}). In some instances the cloud droplet concentration was a substantial fraction of the total number concentration, with some tendency for this fraction to decrease with increasing N_{tot}, whereas in other instances the cloud droplet fraction of the whole was quite small.

Raga and Jonas (1993) compared concentrations of cloud droplets (2 to 47 µm diameter) near the tops of low-updraft-velocity marine cumulus clouds to concentrations of below-cloud accumulation-mode aerosol particles (0.1 to 3 µm diameter), Figure 10. Again a wide range of aerosol particle concentrations was encountered, 50 to 5000 cm^{-3}. The sensitivity of cloud droplet concentration to aerosol particle concentration was much lower than that observed by Twomey and Warner (1967) or by Pueschel et al. (1986).

Martin et al. (1994) compared concentrations of cloud droplets (0.5 to 47 µm diameter) in warm stratocumulus clouds to concentrations of aerosol particles (0.1 to 3 µm diameter) in air just below cloud base (Figure 11). At low aerosol particles characteristic of air relatively free of anthropogenic influence the concentration of cloud drops was nearly equal to the concentration of aerosol particles, whereas there was a marked fall off at higher aerosol particle concentrations ($\gtrsim$ 400 cm^{-3}) characteristic of anthropogenic influence.

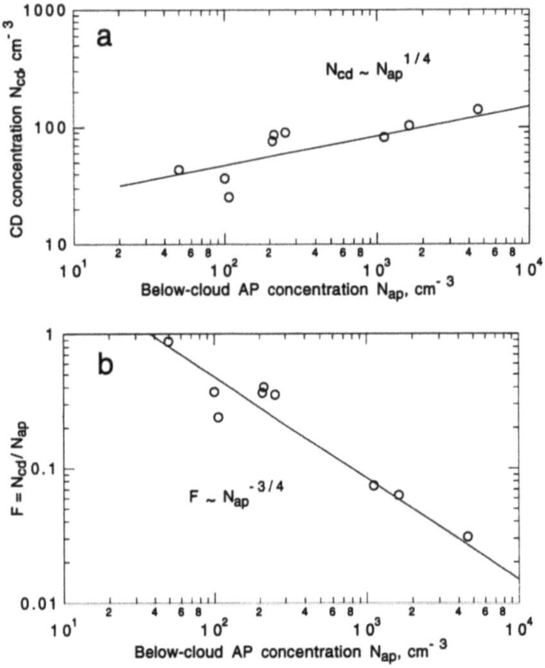

Figure 10. Dependence of cloud droplet concentration N_{cd} on below-cloud accumulation-mode aerosol particle concentration N_{ap} for marine cumulus (data of Raga and Jonas, 1993). Panel (a) shows an approximate 1/4 power dependence of N_{cd} on N_{ap} over a wide range of N_{ap}. Panel (b) shows the corresponding fractional incorporation of aerosol particles approaching unity at low N_{ap} but falling off considerably with increasing N_{ap}.

Gillani et al. (1992 a, b) simultaneously measured concentrations of cloud droplets N_{cd} (2 to 47 μm diameter) and interstitial accumulation mode particles N_{ap} (0.1 to 3 μm diameter) in stratiform clouds in the vicinity of Syracuse New York. Results were reported in terms of the fraction of total particles present as cloud droplets, $F \equiv N_{cd}/N_{tot}$, where $N_{tot} = N_{cd} + N_{ap}$. Examples of time series of F from this study are shown in Figure 12. For the measurements shown in panel (a) F is

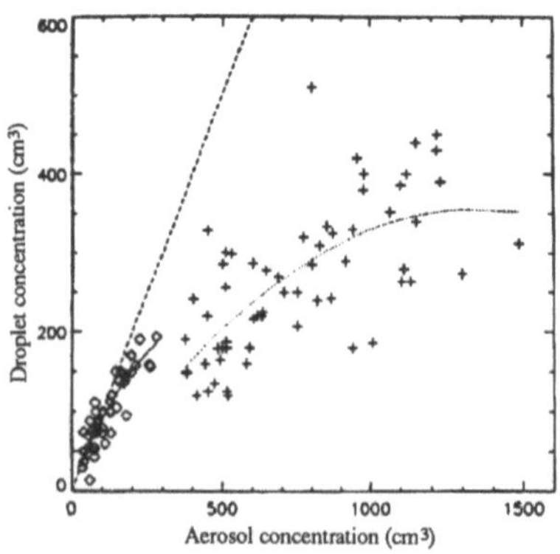

Figure 11. Comparison of concentrations of cloud droplets (0.5 to 47 μm diameter) in warm stratocumulus clouds to the concentrations of aerosol particles (0.1 to 3 μm diameter) in air just below cloud base. ◇, Maritime airmasses; +, continental airmasses. The solid line is the best-fit curve for maritime airmasses, the dotted line the best-fit curve for the continental air masses, and the dashed line is $y = x$. From Martin et al. (1994).

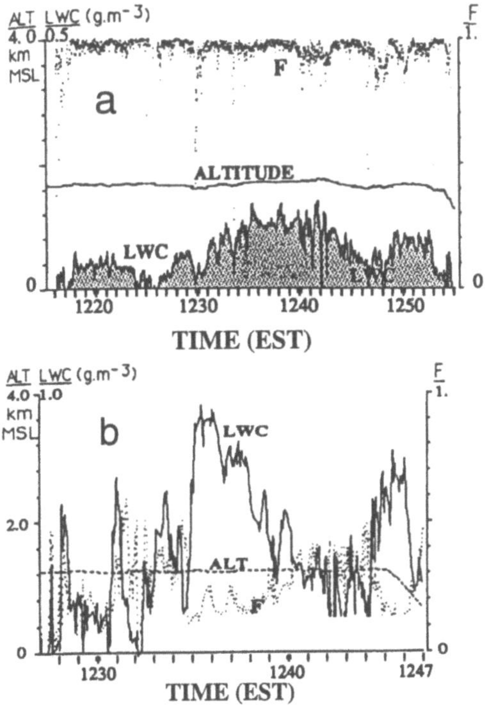

Figure 12. Time series of liquid water content (LWC) and altitude (left ordinates) and fractional incorporation of aerosol particles $F = N_{cd}/(N_{cd} + N_{ap})$ measured during aircraft flights in liquid water stratiform clouds in vicinity of Syracuse, New York on two different dates. Panel (a) (cloud layer 14 in Figure 13) shows instance of uniformly high F, whereas panel (b) (cloud layer 2 in Figure 13) shows instance of low and variable F, both in continuous cloud decks. Gillani et al. (1992a).

near unity throughout the time period shown, whereas for those in panel (b) F is substantially below unity. Figure 13a shows a plot of F against N_{tot} for the 19 cloud layers examined in this study, indicating that F is near unity at low N_{tot} ($\lesssim$ 300 cm^{-3}), but decreases with increasing N_{tot}, suggestive of a possible reduction of maximum supersaturation resulting from competition for water vapor by increased particle concentrations. However at any given value of N_{tot} substantial variability in F was observed, especially at higher values of N_{tot}. Figure 13b shows the measurements of F plotted against mean lapse rate in the cloud taken as a surrogate for updraft velocity, which was not measured, and, in any event, is very difficult to determine. The correlation here is as great as in the plot versus N_{tot}, with F increasing with increasing instability in the cloud layer. It is instructive, to locate particular cloud layers in the two figures. Thus cloud layers 4 and 10, which exhibit highly differing values of F for the same value of N_{tot}, are seen to arise from situations having very different lapse rates. Likewise layers 1 and 3, which seem low in the plot of F vs. lapse rate, are characterized by high particle concentrations. The mutual dependence of F on N_{tot} and vertical velocity

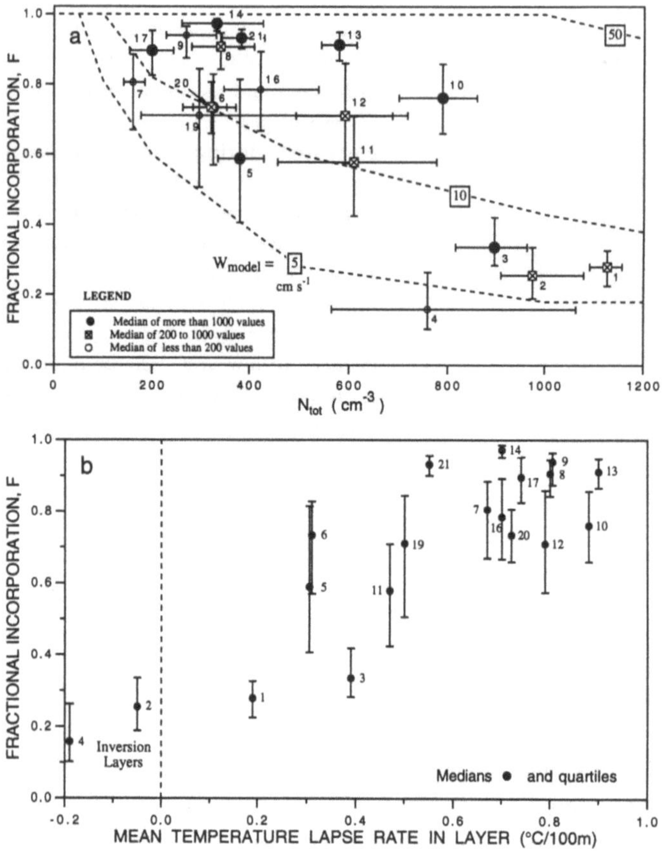

Figure 13. Dependence of $F = N_{cd}/N_{tot}$ on $N_{tot} = N_{cd} + N_{ap}$ (a) and on mean lapse rate in cloud layer (b) in 19 liquid water stratiform cloud layers sampled in the vicinity of Syracuse New York. Panel (a) shows decreasing tendency of F with N_{tot} whereas panel (b) shows sensitivity of F to stability, a surrogate for updraft velocity. Dashed curves in (a) show calculated fractional scavenging efficiency for assumed aerosol particle size distribution as a function of N_{tot} and indicated updraft velocities (cm s^{-1}). Gillani et al. (1992b).

is supported by model calculations shown by the dashed lines in Figure 13a. Thus while such an explanation probably does not account for all the variability in F, including the within-layer variability, it suggests that the observations are at least broadly consistent with the understanding represented in the model.

Another approach to the sensitivity of cloud droplet or CCN concentrations to anthropogenic influences is by direct examination of the dependence of these concentrations to simultaneously measured mass concentrations of a specified substance, such as sulfate or acidity. For reasons indicated above sulfate is expected to exert a dominant influence on cloud droplet concentrations. Several investigators have examined the relation between N_{ccn} or N_{cd} and sulfate concentrations in a variety of locations from industrially influenced to pristine; a

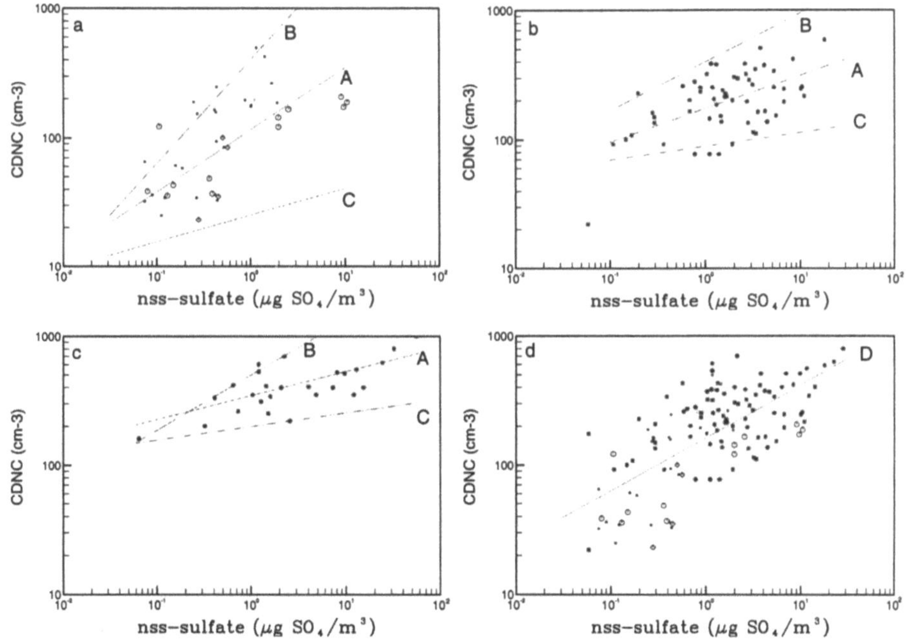

Figure 14. Dependence of cloud droplet or CCN concentration on non-seasalt sulfate aerosol mass concentration for (a) maritime clouds; (b) continental stratiform clouds; (c) continental cumuliform clouds, and (d) all data. The three lines A-C correspond to the mean regression line and the maximum and minimum envelopes, respectively, for each of the data sets. The line D is a regression fit to the entire data set. Sources of the data are: (∗) Leaitch et al. (1992); (◇) Quinn et al. (1993); (○) Hegg et al. (1993); (▪) Berresheim et al. (1993); and (∘) Van Dingenen et al. (1995). From Boucher and Lohmann (1995).

number of such comparisons, classified according to cloud type, have been summarized by Boucher and Lohmann (1995) and are shown in Figure 14: aircraft measurements of cloud droplet concentrations, central Ontario and upper New York state, Leaitch et al. (1992); shipboard measurements of CCN active at 0.3% supersaturation, northeast Pacific Ocean, Quinn et al. (1993); aircraft measurements of CCN active at 1% supersaturation, northeast Atlantic Ocean, Hegg et al. (1993); CCN active at 0.3% supersaturation in maritime air, coastal northwest Washington state, Berresheim et al. (1993); shipboard measurements of aerosol particles that are inferred to have previously been cloud droplet nuclei, North Atlantic Ocean, Van Dingenen et al. (1995). The sensitivities indicated by the slopes on the double logarithmic plots are substantially less than unity,

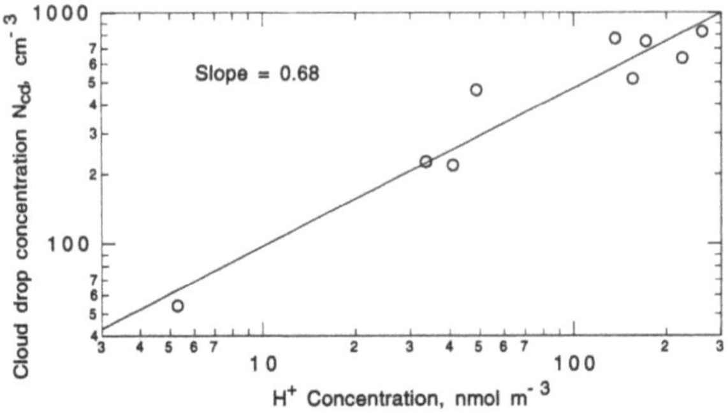

Figure 15. Dependence of cloud droplet concentration N_{cd} on cloudwater acidity (product of H^+ concentration and liquid water content) at Mount Mitchell, North Carolina (elevation 2000 m). Evaluated from data of Grovenstein et al. (1994).

ranging from about 0.2 (continental) to 0.5 (maritime). A similar low sensitivity in maritime air at El Yunque peak, Puerto Rico, (~ 1000 m elevation) is reported by Novakov et al. (1994). In contrast Grovenstein et al. (1994) present measurements at Mt. Mitchell in rural North Carolina (elevation 2000 m) of N_{cd} and cloudwater pH. A double logarithmic plot of N_{cd} versus cloudwater acidity (product of H^+ aqueous concentration and liquid water content) shows a fairly high sensitivity (slope 0.68), Figure 15. The high acidities (pH as low as 3 at liquid water content of 0.2 cm^3 m^{-3}) are unequivocally due to anthropogenic emissions.

In summary, the studies to date can be taken as demonstrating a sensitivity of cloud droplet concentration to the influence of anthropogenic aerosols that is consistent with substantial indirect radiative forcing. However there remains much uncertainty in the quantitative description of the relation between cloud droplet concentration and aerosol loading necessary for accurate modeling of this effect. Clearly any such modeling will have to take into account other variables in addition to aerosol loading, such as local updraft velocity, perhaps including turbulent influences.

<u>Studies relating cloud albedo to cloud droplet concentration</u>. Although there is a convincing theoretical relation between cloud albedo and the size distribution of droplets in the cloud, the phenomenon is difficult to demonstrate because of the inherent variability in cloud albedo due to variability in liquid water path. One potentially convincing approach to quantifying the phenomenon is by means of ship tracks, which afford an immediate contrast to proximate unperturbed regions of the cloud. As noted above, an increase in cloud droplet number concentration and decrease in effective radius have been observed in ship tracks compared to adjacent cloud regions and likewise an increase in cloud optical depth (Radke et al., 1989; King et al., 1993). Unfortunately, from the perspective of

obtaining an unambiguous demonstration of the effect of increased cloud droplet concentration on cloud albedo, the ship tracks reported in this study also exhibited substantially enhanced liquid water content that undoubtedly contributed to the enhancement of cloud albedo associated with the ship tracks. An explanation that has been offered for this is suppression of drizzle resulting from the reduction in cloud droplet size.

A few studies have examined the relation between cloud droplet microphysical properties and cloud albedo, either on short time scales or in longer time averages.

Durkee (1988) presented an image, derived from the 3.7 μm channel of the satellite borne Advanced Very High Resolution Radiometer (AVHRR), of enhancement of reflectivity of marine stratus cloud by the urban plume emanating from San Francisco, California, together with *in-situ* measurements of the cloud droplet mean radius and total aerosol particle concentration. Aerosol particle concentration is enhanced by an order of magnitude in the urban plume. The reflectance closely follows the expected dependence on mean radius in the infrared channel. The lack of a corresponding signal in the visible channel (0.63 μm) was ascribed to confounding variability in the liquid water path.

Kaufman and Nakajima (1993) examined the effect of enhanced aerosol loading on the reflectance (as measured by AVHRR) of low cumulus and stratocumulus clouds over the Brazilian Amazon basin during periods of biomass burning. Smoke loading was inferred from optical depths in clear sky regions proximate to the clouds. A distinct increase was observed in the 3.7 μm reflectivity with increasing aerosol loading. However a similar increase in the visible reflectivity (0.63 μm) was not observed; this was attributed to increased absorption in the clouds due to soot within the aerosol offsetting the brightening influence expected from the increased concentration of cloud droplets.

Grovenstein et al. (1994) presented measurements of cloud albedo, derived from the 0.63 μm channel of the AVHRR, together with in situ measurements of cloud microphysical and chemical properties at Mount Mitchell, North Carolina for four different days in summer 1993. Figure 16 shows cloud reflectivity plotted as a function of the concurrent cloud drop concentration; also shown is the dependence on cloud albedo expected from equation (5b). The agreement between observations and theory is surprisingly good, in view of the constancy of liquid water path assumed in derivation of that equation and the inevitable

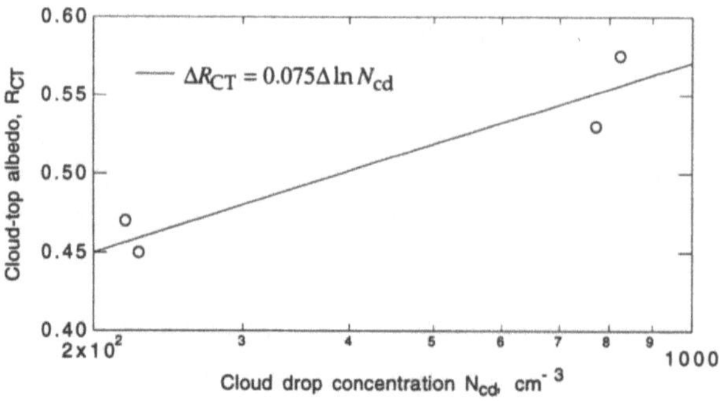

Figure 16. Dependence of cloud albedo at 0.63 μm on cloud drop concentration N_{cd} at Mount Mitchell, North Carolina (elevation 2000 m). Evaluated from data of Grovenstein et al. (1994). The line represents the theoretical slope given by Eq (5b).

variability in liquid water path; the liquid water content of the sampled clouds varied from 0.15 to 0.24 $cm^3\ m^{-3}$. Thus although the apparent agreement with theory is very good, it would seem that many more realizations of measurements such as these would be required to build statistical confidence in this relationship.

Boers et al. (1994) examined the relation between monthly mean CCN concentration in marine boundary layer air at Cape Grim, Tasmania and optical depth of low marine clouds (cloud-top pressure greater 680 hPa) derived from satellite data. The time series of the data are given in Figure 17. Despite considerable scatter a trend of increasing optical depth with increasing CCN concentration is discernible; in panel (c) a linear fit has been drawn through the data plotted as optical depth versus $(CCN)^{1/3}$, as suggested by equation (4). Panel (d) shows the resulting cloud-top albedo calculated by equation (1) from the reported cloud optical depth and from the regression fit. Also shown in that panel is the change in cloud-top albedo evaluated by equation (5). The latter exhibits a much greater dependence on CCN concentration than is exhibited by the observations. The reasons for this departure are not immediately evident. One possibility, however, is the averaging inherent in the measurements of both the CCN concentrations and the cloud optical depths. Moreover, optical depth will be influenced by variations in quantities other than CCN concentrations, notably cloud liquid water path. One may also question the representativeness of the CCN concentrations in describing the droplet concentrations in the clouds. Nonetheless the markedly lower sensitivity exhibited by the observations than would be expected from equation (4), a factor of 3.5 in slope of albedo versus logarithm of CCN concentration, raises a major question regarding estimates of shortwave forcing based on this treatment. Clearly the reasons for this discrepancy need to be understood.

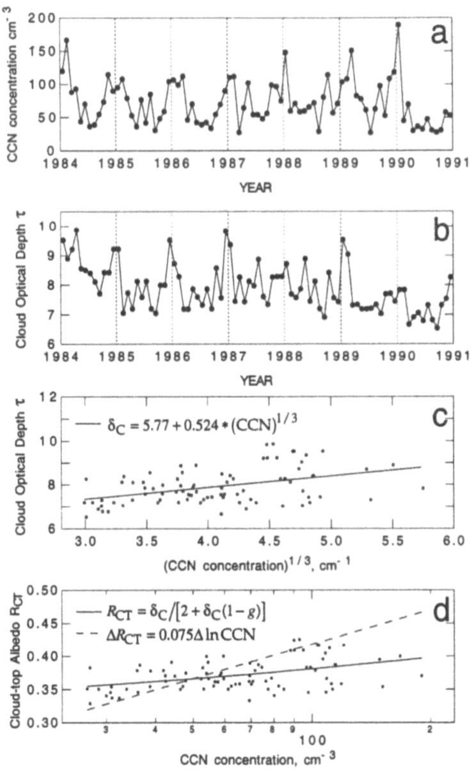

Figure 17. Time series of (a) monthly mean CCN concentration (100m altitude, 0.23% supersaturation) and (b) monthly mean cloud optical depth at Cape Grim, Tasmania; Data of Boers et al. (1994). Cloud optical depths were derived from ISCCP (International Satellite Cloud Climatology Project) data for the nearest grid point to the Cape Grim station that was fully oceanic, and were restricted to cloud-top pressures greater than 680 hPa. Panel (c) shows correlation of optical depth with CCN concentration, where the line represents a linear fit to the data plotted according to Eq (4). Panel (d) shows the corresponding cloud-top albedo evaluated by Eq (1) from the measured optical depth and the corresponding to the regression equation, and also as evaluated by Eq (5b). The slope of the observed dependence of cloud-top albedo on the logarithm of CCN concentration, as represented by the regression fit, is lower by a factor of 3.5 than that given by Eq (5b).

Attempts to identify correlations of satellite observations of albedo with regions influenced by anthropogenic SO_2 emissions have yielded mixed results. Schwartz (1988) attempted to discern interhemispheric differences (northern versus southern) in the cloud contribution to planetary albedo based on measurements from the Nimbus 3 satellite. A large interhemispheric difference might be expected because anthropogenic emissions are disproportionately concentrated (roughly 90%) in the northern hemisphere, although the comparison might well be confounded by other interhemispheric differences such as land masses and ocean circulations. Within the uncertainty of the measurements no indication was discerned of enhanced cloud albedo in the northern hemisphere.

Kim and Cess (1993), examined monthly averaged albedo for low-level marine clouds as obtained from the Earth Radiation Budget Experiment (ERBE) as a function of longitude for regions of the North Pacific and North Atlantic expected to exhibit influence of anthropogenic aerosols, and compared these to regions of the southern hemisphere, taken as a control. Figure 18 shows an example of these results for April, 1985. Significant trends were observed indicative of albedo enhancement over coastal areas affected by anthropogenic aerosols. However in a similar study Falkowski et al. (1992) found that the monthly averaged albedo of low-level marine stratus clouds over the North Atlantic Ocean obtained from ERBE exhibited correlation with ocean productivity inferred from the Coastal Zone Color Scanner chlorophyll concentrations, but did not exhibit systematic decreases eastward from North America as would be expected if cloud albedo were enhanced by anthropogenic aerosol. They concluded that much of the variability in cloud albedo could be accounted for by biogenic emissions of sulfate precursors.

Recently, Han et al. (1994) examined the latitudinal dependence of drop sizes of warm ($T > 273.2$ K) clouds, as determined by an algorithm that uses radiances at 3.7 μm and 10.8 μm determined by the satellite-borne AVHRR radiometer. Four seasonal months were examined, and clouds were distinguished according to whether they were above land or ocean. Figure 19 shows the annual (4-month average) zonal-mean radii derived by this analysis. It is seen first that the effective radius is systematically smaller for continental clouds than for maritime clouds, consistent with the discussion given above. More importantly, with respect to identifying possible anthropogenic influences, there is indication, in both the maritime and continental clouds, of systematically smaller effective

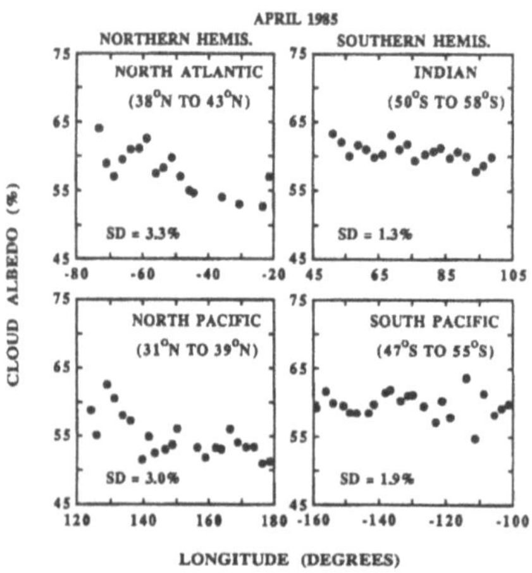

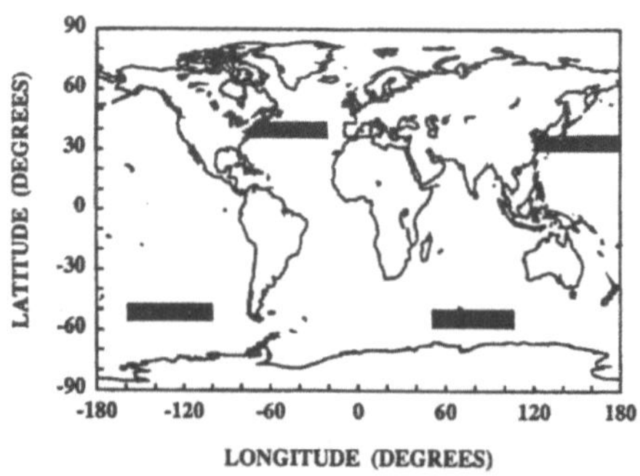

Figure 18. Left panel shows dependence of albedo of low clouds on longitude for four geographical regions indicated in right panel. Data are from April, 1985. From Kim and Cess (1994).

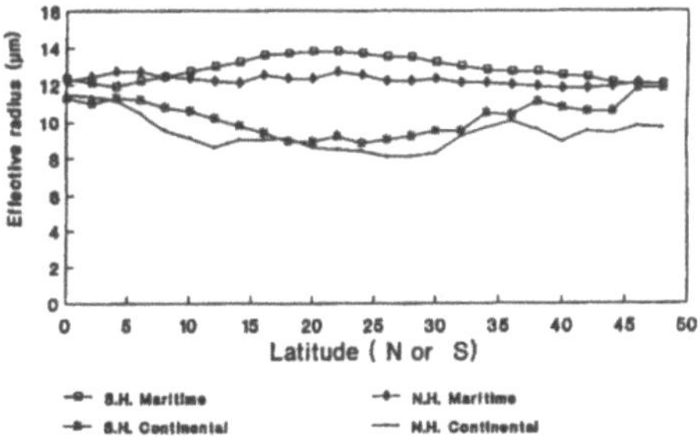

Figure 19. Annual (4-month average) zonal mean water cloud droplet effective radii for 1987 from analysis of NOAA-9 AVHRR radiances. From Han et al. (1994).

radius in mid latitudes of the Northern Hemisphere, consistent with such anthropogenic influence. Han et al. suggest that because of the effect of the inherent variability in liquid water path on cloud albedo, effective radius is a more sensitive indicator of anthropogenic influence than is cloud albedo.

In summary, despite a great sensitivity in cloud albedo that might be expected due to anthropogenic aerosols, there is scant evidence of this influence from satellite measurements, for example over the North Atlantic. A possible reason for this may be that previous searches for this influence have examined long-time averages, in which fluctuations in liquid water path dominate, especially in view of the short temporal and spatial scales which appear to characterize the variability in the anthropogenic aerosol loading (Benkovitz et al., 1994). On the other hand, examination of effective radius, as has been carried out by Han et al., may prove to be a more sensitive indicator of such influence, especially in short-time averages.

STUDIES WITH CLIMATE MODELS

The estimates of the indirect effect discussed at the beginning of this chapter were made with very simple models. These provide initial guidance as to the possible magnitude of the global forcing of the climate system by changes in aerosol concentrations. For a more detailed assessment, especially of the response to the aerosol forcing, it is necessary to use three-dimensional climate models. These have been widely used to study the CO_2 problem and more recently the direct aerosol effect (e.g. Taylor and Penner, 1994; Mitchell et al., 1995). As discussed below, they are now being applied to the indirect effect of sulfate aerosols. While such models allow key aspects of the problem to be addressed, such as the geographical distribution of the forcing, it must be appreciated that even the most detailed of present climate models represent cloud microphysical processes only crudely, if at all. The models are slowly being improved in this respect, but there are fundamental problems in trying to represent such processes in models whose horizontal resolution is no better than a few hundred kilometers. Cloud microphysical processes and the interactions with aerosols must therefore be parametrized, based on more detailed cloud-resolving models and observational data.

Until recently, the modeling groups working on the development of climate models had little interaction with those developing models of atmospheric chemistry. This situation is changing, because of recognition of the need to treat coupled chemistry-climate problems such as anthropogenic perturbations to the carbon dioxide and ozone distributions and the direct and indirect effects of sulfate aerosols. Fully coupled chemistry-climate models, which are under development in several centers, clearly will be powerful tools for studying the indirect effect. To date, however, the published climate model studies of the indirect effect have been much less ambitious and have concentrated on estimating the geographical distribution of the forcing, to extend the single global values from the simple models. As will be seen, the uncertainties in the overall magnitude of the forcing are so large (in marked contrast to the CO_2 problem) that there is much work to be done in this area before considering the feedbacks induced by the forcing.

Given the large uncertainties in the magnitude of the forcing, it might be argued that there is little to be gained from working with three-dimensional climate models, as opposed to the simple models. This might be true if the only parameter of interest was the globally averaged forcing. However, in order to simulate the geographical distribution of the forcing, the climate models must first simulate the geographical distribution of the effective radius r_e. With the advent of global satellite retrievals of r_e such as those of Han et al. (1994), it is now possible to evaluate climate models in ways which are not available for simple models. A climate model which is able to reproduce the observed distributions of the top of the atmosphere radiation budget, as well as the cloud amounts, liquid water contents and r_e, is clearly a much more powerful tool for studying the indirect forcing than a simple model which cannot be evaluated in this way.

The most advanced climate models now calculate the radiative properties of water clouds from the cloud liquid water content and effective radius r_e. The liquid water content is modeled explicitly, using a prognostic equation which takes into account the sources and sinks of cloud water (e.g. Smith, 1990). In contrast, r_e is prescribed, usually as a single global value or perhaps as different values for land and sea, to approximate the different aerosol characteristics of continental and maritime air masses. In order to study the indirect effect, it is necessary to include r_e as an additional model variable by representing its dependence on the liquid water content and CCN concentration, and hence on the aerosol particle concentration.

The physical processes which control r_e are described earlier in the chapter. In a cloud-resolving model these processes may be represented by an explicit treatment of the growth of cloud drops, but given the coarse temporal and spatial resolution of a GCM this approach is neither appropriate nor computationally feasible. Instead, this detail is circumvented by parametrizing the relationships found in field studies between the number of cloud drops and the aerosol concentrations.

Two climate model studies of the indirect forcing have recently been conducted (Jones et al., 1994; Boucher and Lohmann, 1995). Both studies employed externally specified distributions of sulfate aerosol calculated by the MOGUNTIA chemical transport model (Langner and Rodhe, 1991), and from these calculated the distribution of cloud drop concentrations. Given the cloud liquid water content it was possible to derive the effective radius r_e and thence the cloud radiative properties. In both studies, it was assumed that the dominant source of CCN is sulfate aerosol. Comparisons between simulations with the present day sulfate distributions and those for pre-industrial conditions provided the estimates of the indirect forcing, expressed in W m^{-2}.

Jones et al. (1994) used the climate model of the UK Hadley Centre. This model includes the prognostic cloud scheme of Smith (1990), which calculates the cloud amounts and liquid water contents. A control integration of the model was run in order to represent present-day conditions, using the mean sulfate aerosol distribution from the Langner and Rodhe simulations. These distributions are expressed in terms of the aerosol mass loading, so this was transformed to a number concentration by assuming a log-normal size distribution. This was done in order to capitalize on the aircraft data obtained by Martin et al. (1994) for boundary layer stratocumulus clouds in several regions of the globe. Martin et al. showed that there is a good correlation between the number concentration of aerosols in the sub-cloud layer and the number of cloud drops (Figure 11). There is almost a one-to-one correspondence in clean maritime airmasses, but in polluted continental air only a small fraction of the aerosol is activated to become cloud drops. Jones et al. fitted these data with a simple analytic expression, to give the number concentration (in cm^{-3}) of cloud droplets (N_{cd}) from the aerosol concentration A (also in cm^{-3}):

$$N_{cd} = 375(1 - e^{-A/400}) \qquad (12)$$

Since the cloud liquid water content L is a predicted variable of the model, it is then a simple matter to derive the effective radius r_e as

$$r_e = (3L/4\pi\kappa N_{cd})^{1/3} \qquad (13)$$

where κ was defined in equation (4) and in the present context accounts for the different shapes of maritime and continental drop size distributions. Figure 20 shows the annual-mean r_e distribution from the control integration, for present-day aerosol loadings. This is in reasonable agreement with the satellite retrievals of Han et al. (1994), providing some support for the method, although there are differences which need to be studied further.

Jones et al. (1994) then performed a series of single timestep integrations of the climate model, in which the aerosol distribution was replaced by that for natural

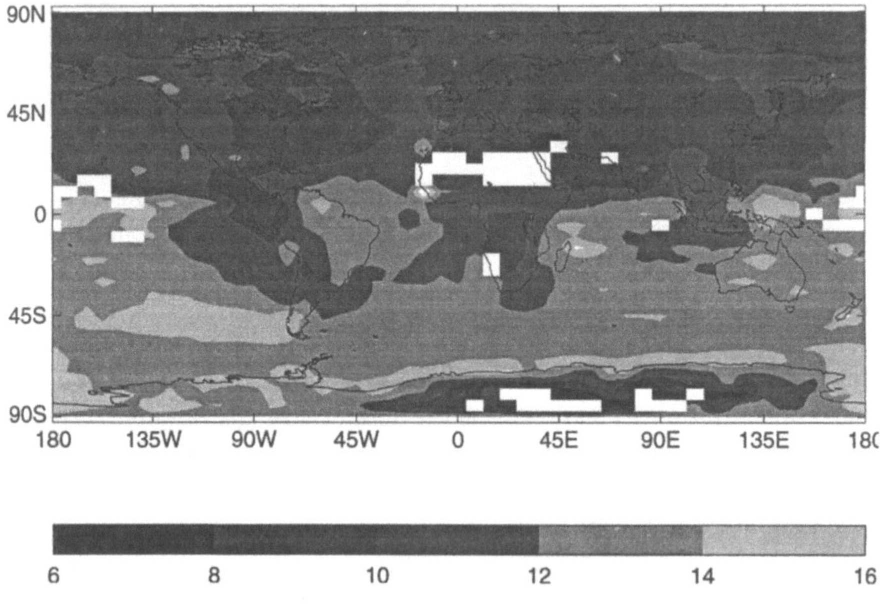

Figure 20. Annual-mean composite of the low cloud droplet effective radius for GCM simulation of the influence of natural and anthropogenic sulfate aerosols on cloud droplet effective radius by Jones et al. (1994). Scale bar indicates droplet radius in μm. Mean radius = 11.5 μm.

sources only. The difference compared with the control simulation provided the estimate of the current contribution from anthropogenic sulfate to the indirect effect. Figure 21a shows the change in r_e due to the anthropogenic sulfate and Figure 21b the corresponding distribution of the associated radiative forcing. The largest increases in r_e, of over 4 μm, are in regions where the anthropogenic contribution to the total sulfate concentrations is largest, such as eastern North America, eastern Europe, and China. These regions are also where the radiative forcing is greatest, peaking at about -5 W m^{-2}. The magnitude of the forcing is also significant in the southern hemisphere, despite the small amount of sulfate pollution. This is due to the high susceptibility of the clouds in such remote regions, where the number concentrations of cloud droplets are low. According to equation (6), this amplifies the sensitivity of the cloud albedo to the sulfate loading and produces a significant radiative perturbation, even for the fairly low incremental sulfate loading. The global average value for the indirect effect obtained by Jones et al. is -1.3 W m^{-2}, although there are considerable uncertainties in this figure. These result from many sources, including the use of the sulfate mass loadings to infer the parameter A, the assumed aerosol size distribution, the generality or otherwise of equations (12) and (13), and the sensitivity of the results to the quality of the climate model simulations.

Boucher and Lohmann (1995) performed a rather similar study, although there are several pertinent differences. They compared results from two climate models: the LMD model developed at the Laboratoire de Météorologie Dynamique in Paris and the ECHAM model of the Max Planck Institute for Meteorology in Hamburg. They derived the droplet number concentrations directly from the aerosol mass loadings, by fitting analytic expressions to data from field studies (Figure 14). This is potentially a good approach, as it circumvents the assumption made by Jones et al. regarding the aerosol size distribution, provided the field studies measure both the sub-cloud sulfate mass loadings and the in-cloud droplet number concentrations. Unfortunately, no such comprehensive measurements have yet been made; for example, Leaitch et al. (1992) made measurements of the cloud water sulfate rather than of the sub-cloud aerosol, and in other field studies summarized in Figure 14 there were no measurements of the droplet concentrations, which instead were assumed to be the same as the number of CCN in the sub-cloud layer. This may account for some of the scatter in Figure 14, and further emphasizes the need for more comprehensive field studies.

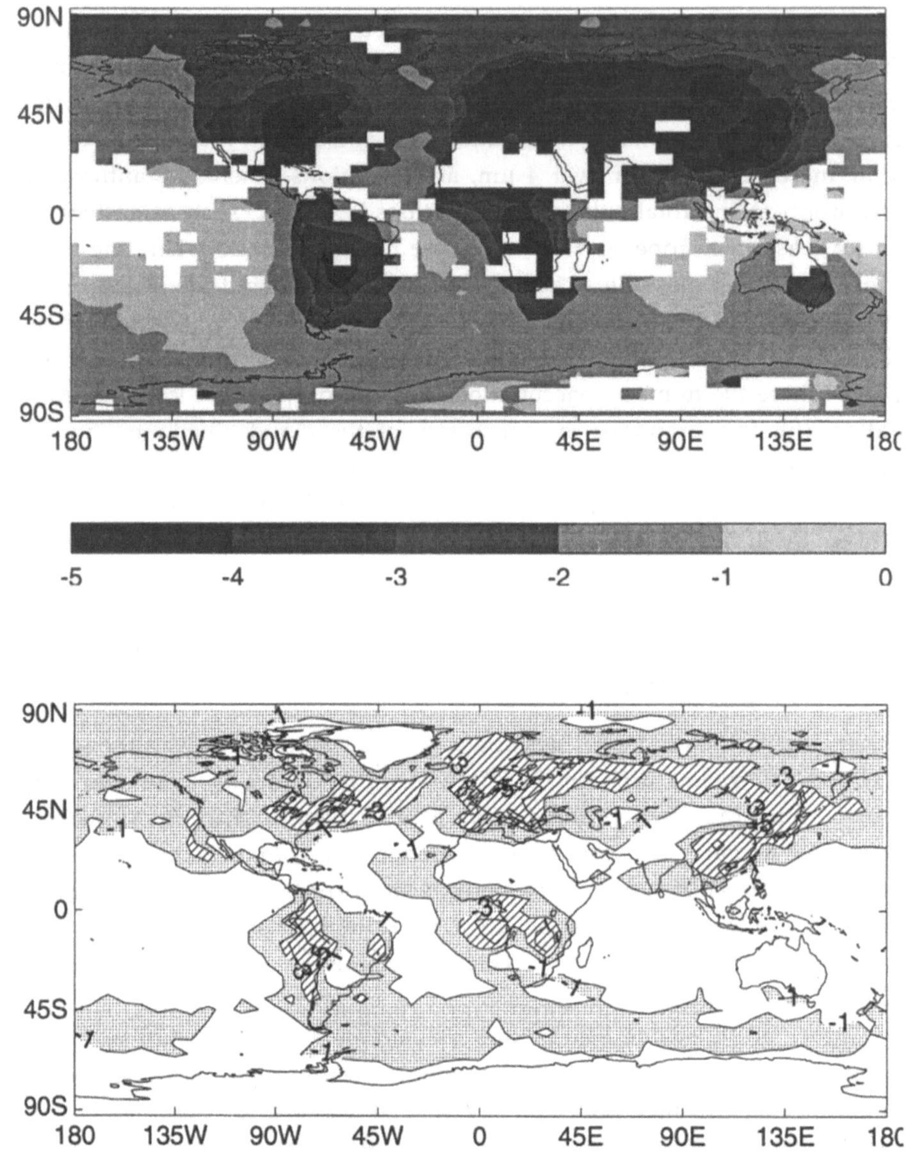

Figure 21. (a) Annual-mean composite of the instantaneous change in low cloud droplet effective radius due to changing from natural-only to total aerosol concentration in the GCM simulation of the influence of sulfate aerosols on cloud droplet effective radius by Jones et al. (1994). Scale bar indicates change in droplet radius in μm. Mean change = -2.2 μm. (b) Distribution of the indirect radiative forcing due to the change in aerosol concentration, in terms of the change in the top of atmosphere (TOA) net radiation. Mean forcing = $-1.3\ W\ m^{-2}$.

Boucher and Lohmann (1995) fitted regression lines to the data for different cloud types, as well as to the dataset as a whole, and also produced maximum and minimum "envelopes" to assign error bars on the calculated forcing (Figure 14). They evaluated the simulations of r_e by comparing with the Han et al. (1994) retrievals, as was done by Jones et al.

Table 3 lists the estimates of the indirect forcings obtained by Boucher and Lohmann (1995) with the two climate models, for four experiments corresponding to the regression lines A-D shown for the several cloud types in Figure 14. For their main experiments (corresponding to regression lines A and D) the results were very consistent, with a global average forcing of about -1.0 W m^{-2}. This is in good agreement with the value obtained by Jones et al. (1994), -1.3 W m^{-2}, although it should be borne in mind that both studies were based on the same modeled distribution of aerosol sulfate loading. Table 3 also shows a range of values from -0.5 to -1.5 W m^{-2}, but, as pointed out by Boucher and Lohmann, even this rather large range probably underestimates the true uncertainty. Figure 22 shows the geographical distributions of the forcing from the two climate models for the experiment corresponding to regression line D. The distributions are very similar to those from the Jones et al. study (Figure 21b). These figures emphasize the complex geographical distribution of the indirect forcing,

Table 3. Indirect radiative forcing by sulfate aerosols, for experiments by Boucher and Lohmann (1994) with the LMD and ECHAM climate models.

		Forcing (W m^{-2})		
Experiment	Model	NH	SH	Globe
A	LMD	-1.6	-0.4	-1.0
	ECHAM	-1.4	-0.6	-1.0
B	LMD	-2.2	-0.6	-1.4
	ECHAM	-2.1	-0.9	-1.5
C	LMD	-0.8	-0.2	-0.5
	ECHAM	-0.6	-0.3	-0.45
D	LMD	-1.7	-0.4	-1.05
	ECHAM	-1.6	-0.6	-1.1

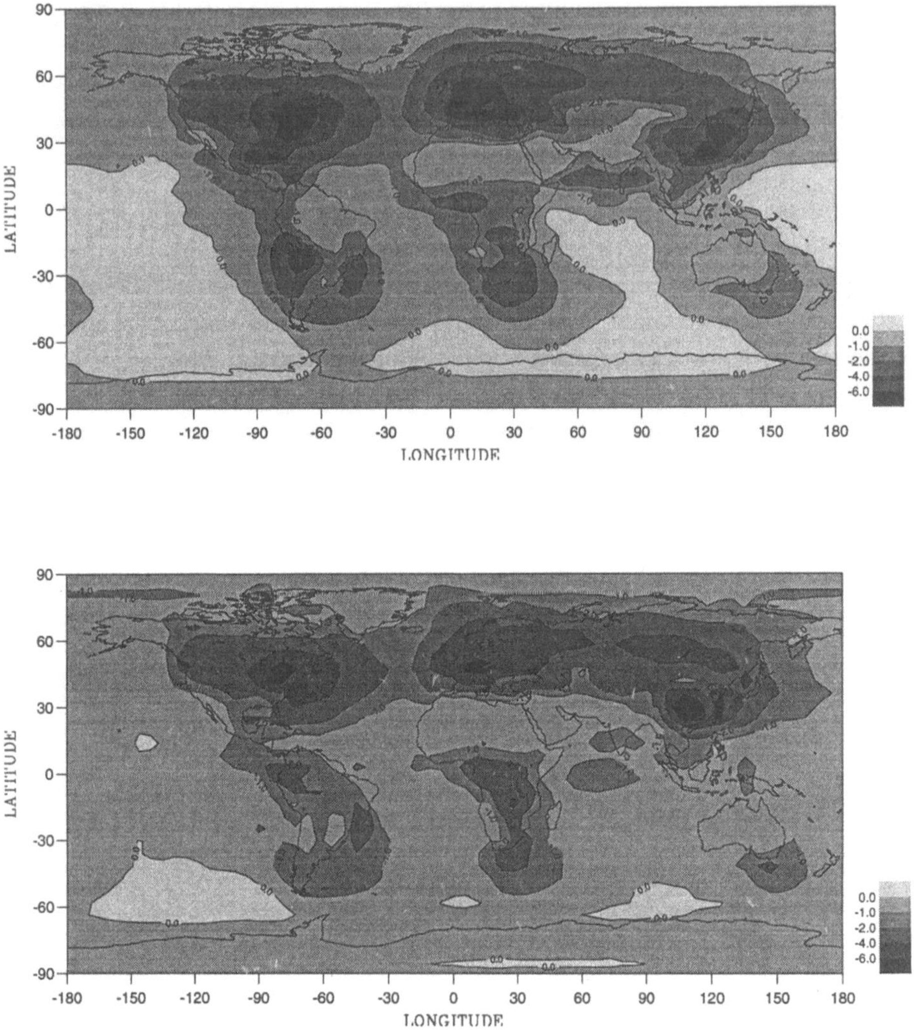

Figure 22. Calculated annual-mean loss of solar irradiance due to the indirect effect of sulfate aerosol for experiment D of Boucher and Lohmann (1995), corresponding the regression D in Figure 14. Top panel gives results for the LMD model and bottom panel for the ECHAM model. Scale bar indicates forcing in W m^{-2}.

compared with the much smoother distribution of the forcing from increases in trace gases (e.g., Kiehl and Briegleb, 1993).

One of many problems with using three-dimensional climate models to estimate the indirect effect is the difficulty in producing a realistic simulation of r_e.

Boucher (1995) attempted to circumvent this problem by constraining the LMD model with the retrieved values of effective radius obtained by Han et al. (1994). The indirect effect was then estimated from the contrast in r_e, and hence in the radiation budget, between the two hemispheres. This is a more direct version of the approach employed by Schwartz (1988). Boucher obtained a value of -0.6 to -1.0 W m^{-2} for the indirect forcing in the latitude band 0 to 50° N, in broad agreement with the climate model estimates.

DISCUSSION AND CONCLUSIONS

The Jones et al. (1994) and Boucher and Lohmann (1995) studies represent first attempts to study the indirect forcing issue with climate models, using externally specified aerosol fields. There are several directions in which this work may be extended.

There is much that needs to be done to reduce uncertainties in the forcing and in its distribution. Both climate model studies used annual-mean sulfate aerosol distributions from the MOGUNTIA model. There is a considerable seasonal cycle in the sulfate concentrations, which needs to be included in the models. Simulations from other chemistry models also need to be included for comparison. Both climate model studies assumed that only sulfate aerosols grow to become cloud drops, but other aerosols are known to be sources of CCN and should be represented. Some of the disagreements between the r_e distributions from the models and the satellite retrievals of Han et al. (1994) might be caused by this simplification. More detailed comparisons with the satellite data are clearly necessary, as these provide a crucial test of the models. Other methods for predicting the droplet concentrations should also be investigated; this will help to refine the uncertainty limits explored by Boucher and Lohmann. For example, Ghan et al. (1993) have developed a method based on the cloud-scale vertical velocity, although it is a considerable challenge to derive this from the grid-scale variables.

Despite the large uncertainties in the magnitude of the forcing, it is desirable to integrate the climate models forward to study the feedbacks induced by the forcing. One of the first feedbacks to be considered should be that suggested by Albrecht (1989), whereby increased aerosol concentrations not only brighten the clouds but also suppress the important drizzle mode in boundary layer clouds and so increase the cloud lifetime and areal coverage. Modeling this process requires an explicit link between the change in droplet number concentration and the formulation of precipitation processes in the model.

A further step will be to include estimates of the indirect forcing in coupled model simulations of the climate change which has occurred since the industrial revolution and which may occur over the next century. These simulations are performed with coupled ocean-atmosphere models and have only recently included crude representations of the direct aerosol effect. Mitchell et al. (1995) show results which indicate that including the direct aerosol effect in the Hadley Centre coupled model leads to an improvement in the simulation of the past temperature record. It would be valuable to extend such work to include the indirect forcing, although the magnitude is of course even more uncertain than that from the direct effect. Nevertheless, it is important to assess the interaction between both aspects of the sulfate-induced cooling and the greenhouse gas warming, as this may be fundamental to understanding the temperature record over the last century and in making climate predictions for the future.

As noted earlier, several groups are developing coupled chemistry-aerosol-climate models. These will remove the need to specify the sulfate aerosol fields and allow full interaction with the rest of the climate model. Such coupled models will instead specify the geographical distribution of the emissions, allow chemical species and aerosols to be transported by the model winds and removed by sink processes such as dry and wet deposition. There are considerable computational challenges in such developments, but they should enable a much more comprehensive examination of the effects of aerosol on climate than has been possible hitherto. In particular, they will allow the models to represent better the extremely heterogeneous nature of the aerosol distribution in both space and time. The studies performed so far with climate models have used temporally averaged aerosol fields and so fail to represent the episodic nature of the indirect forcing, as the pollution is distributed unevenly by the meteorological fields.

All modeling relies on observations made in the real world to develop parametrizations and evaluate simulations. Comprehensive field experiments are needed which address the many uncertain links in the long chain of interactions leading from the production and transport of the aerosol through to the cloud droplet size and radiative properties. No previous field experiment has studied this problem with sufficient breadth, but such experiments are planned. Of particular concern is the large amount of scatter on Figures 11 and 14, which suggests that the relationship between sulfate mass and cloud droplet number concentration is more complex than assumed in the parametrizations used so far. Future field studies should aim to determine the factors which lead to such large scatter, so that more physically based parametrizations can be developed.

Finally, the importance of satellite retrievals of r_e needs to be emphasized again. The retrievals performed by Han et al. (1994) are of great value in assessing the climate model simulations discussed earlier, but much more work is needed to extend the retrievals and to assess their accuracy. Of particular interest is the significant land-sea contrast in these results, the magnitude of which was not simulated well by any of the models. Further retrievals using different methods and data would enable a more comprehensive study of this result. Up to now only time-means of these retrievals have been studied, but it would also be valuable to use the data to perform case studies of aerosol modification of cloud properties and thereby validate the relationship between aerosols and particle size.

Acknowledgments

Work by SES was supported by the Environmental Sciences Division of the U.S. Department of Energy (DOE) as part of the Atmospheric Radiation Measurement Program and was performed under the auspices of DOE under Contract No. DE-AC02-76CH00016. Work by AS was supported by the UK Department of the Environment under contract PECD 7/12/37.

REFERENCES

Ackerman A. S., Toon O. B. and Hobbs P. V. (1993) Dissipation of marine stratiform clouds and collapse of the marine boundary layer due to the depletion of cloud condensation nuclei by clouds. *Science* **262**, 226-229.

Ahr M., I. Flossmann A. and R. Pruppacher H. (1989) A comparison between two formulations for nucleation scavenging. *Beitr. Phys. Atmosph.* **62**, 321-326.

Albrecht B. A. (1989) Aerosols, cloud microphysics, and fractional cloudiness. *Science* **245**, 1227-1230.

Alkezweeny A. J., A. Burrows D. and A. Grainger C. (1993) Measurements of cloud-droplet size distributions in polluted and unpolluted stratiform clouds. *J. Appl. Meteorol.* **32**, 106-115.

Andreae M. O. (1995) Climate effects of changing atmospheric aerosol levels. In *World Survey of Climatology, Vol. XX, Future Climate of the World* (edited by Henderson-Sellers A.). In press.

Ayers G. P. and Gras J. L. (1991) Seasonal relationship between cloud condensation nuclei and aerosol methanesulphonate in marine air. *Nature* **353**, 834-835.

Benkovitz C. M., Berkowitz C. M., Easter R. C., Nemesure S., Wagener R. and Schwartz S. E. (1994) Sulfate over the North Atlantic and adjacent continental regions: Evaluation for October and November 1986 using a three-dimensional model driven by observation-derived meteorology. *J. Geophys. Res.* **99**, 20725-20756.

Berresheim H., Eisele F. L., Tanner D. J., McInnes L. M., Ramsey-Bell D. C. and Covert D. S. (1993) Atmospheric sulfur chemistry and cloud condensation nuclei (CCN) concentrations over the northeastern Pacific coast. *J. Geophys Res.* **98**, 12701-12711.

Boers R., Ayers G. P. and Gras G. L. (1994) Coherence between seasonal variation in satellite-drived cloud optical depth and boundary layer CCN concentration at a mid-latitude Southern Hemisphere station. *Tellus* **46B**, 123-131.

Bohren C. F. (1987) Multiple scattering of light and some of its observable consequences. *Am. J. Phys.* **55**, 524-533.

Boucher O. (1995) GCM estimate of the indirect aerosol forcing using satellite-retrieved cloud droplet effective radii. *J. Clim.* **8**, 1403-1409.

Boucher O. and Lohmann U. (1995) The sulfate-CCN-cloud albedo effect: A sensitivity study with two general circulation models. *Tellus* (in press).

Braham R. R. Jr. (1974) Cloud physics of urban weather modification--A preliminary report. *Bull. Amer. Meteor. Soc.* **55**, 100-106.

Charlson R. J. and Heintzenberg J., eds. (1995) *Aerosol Forcing of Climate.* Wiley, Chichester. In Press.

Charlson R. J., Lovelock J. E., Andreae M. O. and Warren S. G. (1987) Oceanic phytoplankton, atmospheric sulphur, cloud albedo and climate. *Nature* **326**, 655-661.

Charlson R. J., Lovelock J. E., Andreae M. O. and Warren S. G. (1989) Sulphate aerosols and climate. *Nature* **340**, 437-438.

Charlson R. J., Langner J., Rodhe H., Leovy C. B. and Warren S. G. (1991) Perturbation of the Northern Hemisphere radiative balance by backscattering from anthropogenic aerosols. *Tellus* **43AB**, 152-163.

Charlson R. J., Schwartz S. E., Hales J. M., Cess R. D., Coakley J. A. Jr., Hansen J. E. and Hofmann D. J. (1992) Climate forcing by anthropogenic aerosols. *Science* **255**, 423-430.

Coakley J. A. Jr., Bernstein R. L., and Durkee P. A. (1987) Effect of ship-stack effluents on cloud reflectivity. *Science* **237**, 1020-1022.

Durkee P. A. (1988) Observations of aerosol-cloud interactions in satellite-detected visible and near-infrared radiance. In *Preprints, Symposium on the Role of Clouds in Atmospheric Chemistry and Global Climate*, pp. 157-160, American Meteorological Society, Boston.

Falkowski P., Kim Y., Kolber Z., Wilson C., Wirick C. and Cess R. (1992) Natural versus anthropogenic factors affecting low-level cloud albedo over the North Atlantic. *Science* **256**, 1311-1313.

Fitzgerald J. W. (1991) Marine aerosols: A review. *Atmos. Environ.* **25A**, 533-545.

Fitzgerald J.W. and Spyers-Duran P. A. (1973) Changes in cloud nucleus concentration and cloud droplet size distribution associated with pollution from St. Louis. *J. Appl. Meteorol.* **12**, 511-516.

Flossmann A. I., Hall W. D. and Pruppacher H. R. (1985) A theoretical study of the wet removal of atmospheric pollutants--Part I. The redistribution of aerosol particles captured through nucleation and impaction scavenging by growing cloud drops. *J. Atmos. Sci.* **42**, 583-606.

Frisbie P. R. and Hudson J. G. (1993) Urban cloud condensation nuclei spectral flux. *J. Appl. Meterol.* **32**, 666-676.

Ghan S. J., Chuang C. C. and Penner J. E. (1993) A parameterization of cloud droplet nucleation. Part I. Single aerosol type. *Atmos. Res.* **30**, 198-221.

Gillani N. V., Daum P. H., Schwartz S. E., Leaitch W. R., Strapp J. W. and Isaac G. A. (1992a) Fractional activation of accumulation-mode particles in warm continental stratiform clouds. In *Proceedings of the Fifth International Conference on Precipitation Scavenging and Atmospheric-Surface Exchange Processes,* (edited by Schwartz S. E. and Slinn W. G. N.) pp. 345-358, Hemisphere Publishers, Inc., Washington, DC.

Gillani N. V., Schwartz S. E., Daum P. H., Leaitch W. R., Strapp J. W. and Isaac G. A. (1992b) Fractional activation of accumulation-mode aerosols in continental stratiform clouds. WMO Workshop on Cloud Microphysics and Global Change, Aug., Toronto, Ontario, Canada.

Grovenstein J. D., K. Saxena V. and A. Durkee P. (1994) Impact of anthropogenic and natural aerosols on cloud albedo. *Eos. Trans. Geophys. Union* **75** (No. 16, Suppl.), 73.

Hallberg A., Ogren J. A., Noone K. J., Okada K., Heintzenberg J. and Svenningsson I. B. (1994) The influence of aerosol particle composition on cloud droplet formation. *J. Atmos. Chem.* **19**, 153-171.

Han Q., Rossow W. B. and Lacis A. A. (1994) Near-global survey of effective droplet radii in liquid water clouds using ISCCP data. *J. Climate* **7**, 465-497.

Hänel G. (1987) The role of aerosol properties during the condensational growth of cloud: A reinvestigation of numerics and microphysics. *Beitr. Phys. Atmosph.* **60**, 321-339.

Hansson H. C. and Svenningsson B. (1994) Aerosols and clouds. In *Physico-Chemical Behaviour of Atmospheric Pollutants: Proceedings of the Sixth European Symposium Varese, Italy, Oct. 18-22, 1993,* (edited by Angeletti G. and Restelli G.), Vol 2, pp. 837-846. Office of Official Publications of the European Commisssion, Luxembourg.

Hartmann D. L. (1993) Radiative effects of clouds on Earth's climate. In *Aerosol–Cloud–Climate Interactions* (edited by Hobbs P. V.), pp. 151-173. Academic Press, San Diego.

Hegg D. A. (1990) Heterogeneous production of cloud condensation nuclei in the marine atmosphere. *Geophys. Res. Lett.* **17**, 2165-2168.

Hegg D. A., Ferek R. J. and Hobbs P. V. (1993) Light scattering and cloud condensation nucleus activity of sulfate aerosol measured over the Northeast Atlantic Ocean. *J. Geophys. Res.* **98**, 14887-14894.

Hobbs P. V. (1993) Aerosol–Cloud Interactions. In *Aerosol–Cloud–Climate Interactions* (edited by Hobbs P. V.), pp. 33-73. Academic Press, San Diego.

Hudson J. G. (1991) Observations of anthropogenic cloud condensation nuclei. *Atmos. Environ.* **25A**, 2449-2455.

IPCC (1995). Intergovernmental Panel on Climate Change. Radiative Forcing of Climate Change. In *Climate Change 1994* (edited by Houghton J. and Meira Filho L. G.) pp. 1-231. Cambridge University Press, Cambridge.

Jensen J. B. and Charlson R. J. (1984) On the efficiency of nucleation scavenging. *Tellus* **36B**, 367-375.

Jones A., Roberts D. L. and Slingo A. (1994) A climate model study of indirect radiative forcing by anthropogenic sulphate aerosols. *Nature* **370**, 450-453.

Kaufman Y.J., Fraser R.S. and Mahoney R.L. (1991) Fossil fuel and biomass burning effect on climate--heating or cooling? *J. Climate* **4**, 578-588.

Kaufman Y. J. and Nakajima T. (1993) Effect of Amazon smoke on cloud microphysics and albedo. *J. Appl. Meteorl.* **32**, 729-744.

Kaufman Y. J. and Tanré D. (1994) Effect of variations in supersaturation on the formation of cloud condensation nuclei. *Nature* **369**, 45-48.

Kiehl J. T. and Briegleb B. P. (1993) The relative roles of sulfate aerosols and greenhouse gases in climate forcing. *Science* **260**, 311-314.

Kim Y. and Cess R. D. (1993) Effect of anthropogenic sulfate aerosols on low-level cloud albedo over oceans. *J. Geophys. Res.* **98**, 14883-14885.

King M. D., F. Radke L. and V. Hobbs P. (1993) Optical properties of marine stratocumulus clouds modified by ships. *J. Geophys. Res.* **98**, 2729-2739.

Kulmala M., A. Laaksonen, P. Korhonen, T. Vesala and T. Ahonen. (1993) The effect of atmospheric nitric acid vapor on cloud condensation nucleus activation. *J. Geophys Res.* **98**, 22949-22958.

Langner J. and Rodhe H. (1991) A global three-dimensional model of the tropospheric sulfur cycle. *J. Atmos. Chem.* **13**, 225-263.

Leaitch W. R., Isaac G. A., Strapp J. W., Banic C. M. and Wiebe H. A. (1992) The relationship between cloud droplet number concentrations and anthropogenic pollution: Observations and climatic implications. *J. Geophys Res.* **97**, 2463-2474.

Lelieveld J. and Heintzenberg J. (1992) Sulfate cooling effect on climate through in-cloud oxidation of anthropogenic SO_2. *Science* **258**, 117-120.

Lin X., Chameides W. L., Kiang C. S., Stelson A. W. and Berresheim H. (1992) A model study of the formation of cloud condensation nuclei in remote marine areas. *J. Geophys Res.* **97**, 18161-18171.

Martin G., M., Johnson D. W. and Spice A. (1994) The measurement and parameterization of effective radius of droplets in warm stratiform clouds. *J. Atmos. Sci.* **51**, 1823-1842.

Mészáros E. (1992) Structure of continental clouds before the industrial era: A mystery to be solved. *Atmos. Environ.* **26A**, 2469-2470.

Mitchell J. F. B., Johns T. C., J. M. Gregory and S. F. B. Tett. (1995) Transient climate response to increasing sulphate aerosols and greenhouse gases. *Nature* (in press).

Novakov T. and Penner J. E. (1993) Large contribution of organic aerosols to cloud-condensation nuclei concentrations. *Nature* **365**, 823-826.

Novakov T., Rivera-Carpio C., Penner J. E. and Rogers C. F. (1994) The effect of anthropogenic sulfate aerosols on marine cloud droplet concentrations. *Tellus* **46B**, 132-141.

Penner J. E., Charlson R. J., Hales J. M., Laulainen N., Leifer R., Novakov T., Ogren J., Radke L. F., Schwartz S. E. and Travis L. (1994) Quantifying and minimizing uncertainty of climate forcing by anthropogenic aerosols. *Bull. Amer. Meteorol. Soc.* **75**, 375-400.

Platnick S. E. and Twomey S. (1994) Determining the susceptibility of cloud albedo to changes in droplet concentration with the Advanced Very High Resolution Radiometer. *J. Appl. Meteorl.* **33**, 334-347.

Pruppacher H. R. and Klett J. D. (1980) *Microphysics of Clouds and Precipitation.* D. Reidel, Hingham, MA.

Pueschel R. F., Van Valin C. C., Castillo R. G., Kadlecek J. A. and Ganor E. (1986) Aerosols in polluted versus nonpolluted air masses: Long-range transport and effects on clouds. *J. Clim. Appl. Meteor.* **25**, 1908-1917.

Quinn P. K., Covert D. S., Bates T. S., Kapustin V. N., Ramsey-Bell D. C. and McInnes L. M. Dimethylsulfide/cloud condensation nuclei/climate system: Relevant size-resolved measurements of the chemical and physial properties of the atmospheric aerosol particles. *J. Geophys. Res.* **98**, 10411-10427.

Radke L. F. and Hobbs P. V. (1976) Cloud condensation nuclei on the Atlantic seaboard of the United States. *Science* **193**, 999-1002.

Radke L. F., Coakley J. A. Jr., and King M. D. (1989) Direct and remote sensing observations of the effects of ships on clouds. *Science* **246**, 1146-1149.

Raes F. (1995) Entrainment of free tropospheric aerosols as a regulating mechanism for cloud condensation nuclei in the remote marine boundary layer. *J. Geophys. Res.* **100**, 2893-2903.

Raga G. B. and Jonas P. R. (1993) On the link between cloud-top radiative properties and sub-cloud aerosol concentrations. *Quart. J. Roy. Met. Soc.* **119**, 1419-1425.

Savoie D. L. and Prospero J. M. (1989) Comparison of oceanic and continental sources of non-sea-salt sulphate over the Pacific Ocean. *Nature* **339**, 685-687.

SCEP (Study of Critical Environmental Problems). (1970) *Man's Impact on the Global Environment.* MIT Press, Cambridge, MA, 319 pp.

Scorer R. S. (1987) Ship trails. *Atmos. Environ.* **21**, 1417-1425.

Schwartz S. E. (1988) Are global cloud albedo and climate controlled by marine phytoplankton? *Nature* **336**, 441-445.

Shine K. P., Derwent R. G., Wuebbles D. J. and Morcrette J.-J. (1990) Radiative forcing of climate. In *Climate Change: The IPCC Scientific Assessment* (edited by Houghton J. T., Jenkins G. J. and Ephraums J. J.), pp. 41-68. Cambridge Univ. Press, Cambridge.

Slinn W. G. N. (1992) Structure of continental clouds before the industrial era: A mystery to be solved. *Atmos. Environ.* **26A**, 2471-2473.

Smith R. N. B. (1990) A scheme for predicting layer clouds and their water content in a general circulation model. *Quart. J. Roy. Meteor. Soc.* **116**, 435-460.

Taylor J. P. and A. McHaffie. (1994) Measurements of cloud susceptibility. *J. Atmos. Sci.* **51**, 1298-1306.

Taylor K. E. and Penner J. E. (1994) Response of the climate system to atmospheric aerosols and greenhouse gases. *Nature* **369**, 734-737.

Twomey S. (1974) Pollution and the planetary albedo. *Atmos. Environ.* **8**, 1251-1256.

Twomey S. (1977a) *Atmospheric Aerosols.* Elsevier, New York.

Twomey S. (1977b) The influence of pollution on the short-wave albedo of clouds, *J. Atmos. Sci.* **34**, 1149-1152.

Twomey S. (1991) Aerosols, clouds and radiation. *Atmos. Environ.* **25A**, 2435-2442.

Twomey S. and Warner J. (1967) Comparison of cloud droplets and cloud nuclei. *J. Atmos. Sci.* **24**, 702-703.

Twomey S., A. Davidson K. and Seton K. J. (1978) Results of five years' observations of cloud nucleus concentration at Robertson, New South Wales. *J. Atmos. Sci.* **35**, 650-656.

Van Dingenen R., Raes F., and Jensen N. R. (1995) Evidence for anthropogenic impact on number concentration and sulphate content of cloud-processed aerosol particles over the North Atlantic. *J. Geophys. Res.* (in press).

Warner J. and Twomey S. (1967) The production of cloud nuclei by cane fires and the effect on cloud droplet concentration. *J. Atmos. Sci.* **24**, 704-706.

Wigley T. M. L. (1994) Outlook becoming hazier. *Nature* **369**, 709-710.

Wurzler S., Flossmann A. I., Pruppacher H. R. and Schwartz S. E. (1995) The scavenging of nitrate by clouds and precipitation. I. A theoretical study of the uptake and redistribution of $NaNO_3$ particles and HNO_3 gas by growing cloud drops using an entraining air parcel model. *J. Atmos. Chem.* (in press).

SATELLITE OBSERVATIONS OF UPPER TROPOSPHERIC AEROSOLS

G.S. Kent
Science and Technology Corporation, Hampton,
Virginia 23666-0390, USA

Introduction

Many fewer upper tropospheric aerosol measurements have been made than lower tropospheric aerosol measurements. This distinction is particularly true for those that extend over a wide latitude range or that have been continued over a long time period. Moreover, since the concentration of aerosols in the upper troposphere is normally lower than that in the lower troposphere or stratosphere, integral measurements of aerosol optical characteristics made from space reveal very little about the upper troposphere. Some exceptions to these generalizations exist. In the Global Backscatter Experiment (GLOBE) program, developed by the National Aeronautics and Space Administration (NASA) to provide data on the global properties of upper tropospheric aerosols, flight programs were conducted over the Pacific Ocean in which the aerosol properties were measured by *in situ* sampling and remote optical techniques. Some of the data obtained during these flight programs have recently been published (Clarke, 1993; Pueschel et al., 1994). Another considerable set of data on the upper troposphere is that of Hofmann (1993) who has presented the results of 20 years of balloonborne measurements made by optical particle counter over Laramie, Wyoming. Upper tropospheric mass concentrations and optical depths and their long term and seasonal variations are described. Lidar measurements, made using visible or near infrared wavelengths, have yielded a great deal of information about aerosol characteristics in the stratosphere and lower troposphere; however, they can normally only provide similar information about the upper troposphere when there is a particularly strong injection of aerosol from either above or below. Lidar measurements made at CO_2 laser wavelengths (Post, 1986; Post et al., 1992) can more readily distinguish the upper tropospheric aerosols from the molecular background and have the potential of yielding much useful information about these aerosols.

NATO ASI Series, Vol. I 35
Clouds, Chemistry and Climate
Edited by Paul J. Crutzen and Veerabhadran Ramanathan

One set of satellite data that has the capability of providing information about upper tropospheric aerosols is obtained by the solar occultation satellite instruments, the Stratospheric Aerosol and Gas Experiment (SAGE) I, the SAGE II, and the Stratospheric Aerosols Measurement (SAM) II. Several previous publications (Kent et al., 1988, 1991; Kent and Schaffner, 1989) have described the global climatology of the upper tropospheric aerosol in terms of the extinction that it produces at a wavelength of 1 μm. These works have shown more aerosol to lie in the Northern Hemisphere upper troposphere than in the Southern, and that each hemisphere has a similar seasonal variation in aerosol extinction. It was also shown that, away from low latitudes, volcanic aerosol transferred downward from the stratosphere is readily visible. Recently, this analysis has been extended using data from the SAGE II satellite only (Kent et al., 1995). The changes in the aerosol concentration have been studied in more detail and quantified, particularly in terms of the seasonal and latitudinal variations, and the influences of volcanic- and surface-derived aerosol have been separated. In additon, estimates have been made of aerosol size and mass density. A summary of these results will be presented here.

SAGE II Measurements

The SAGE II satellite is the most recent in the SAM II, SAGE I, and SAGE II instrument series that contains sun photometers designed to measure the extinction due to stratospheric aerosols and gases (McCormick et al., 1987). The instrument, which is self-calibrating, measures atmospheric attenuation over a tangential limb path approximately 30 times per day at satellite sunrise and sunset. The multiwavelength data are inverted, under the assumption of horizontal homogeneity, through successive atmospheric layers to give a vertical extinction profile (Chu et al., 1989) with a vertical resolution of 1.0 km. Each derived extinction profile represents an average over a horizontal path approximately 200 km in length. Because of the length of this path the signal received at the satellite instrument is very sensitive to low values of extinction, and small departures from the molecular background can be measured. In the stratosphere, the attenuation due to aerosols is measured at four wavelengths. In the troposphere measurements are made at the two longest wavelengths only (1.02 and 0.525 μm) and then only above an altitude of 6 km. The SAGE II instrument makes successive sunrise or sunset observations at approximately the same latitude and at longitude intervals of about 24°. The latitude of either sunrise or sunset moves from one extreme of latitude to the other over a period of about 1 month (total latitude coverage over this period is about 120°). The instrument thus excels for the study of climatological changes but is less well suited for the observation of fine structure in either time or space. Measurements have been made continually since the satellite launch in October 1984.

In the stratosphere, signal attenuation is produced by atmospheric aerosols and gases only. In the troposphere, the possibility exists, and at lower altitudes the likelihood, that cloud may occur along the optical path from the sun to the satellite instrument. If the cloud is thick it will cause the signal received at the satellite to fall below the threshold of detection. If the cloud is thin (subvisual or near subvisual cirrus) there is the necessity of distinguishing between an event for which no cloud occurs at a given altitude and the attenuation is that due to aerosol only, and one in which cloud is present. This characterization may be done by comparing the extinction at the two primary aerosol wavelengths, since the wavelength dependence of the attenuation due to cloud particles will be different from that due to all but the largest aerosol particles. A model has been developed to apply to the data in order to discriminate between the different physical situations (Kent and McCormick, 1991; Kent et al., 1993). This model has been validated recently by comparison of satellite measurements with lidar and visual observations made along the same optical path. In the analysis presented in the following sections the model has been applied to the satellite data in order to select only those unaffected by cloud along the optical path.

Aerosols in the Upper Troposphere and Lower Stratosphere

A time sequence of the 1.02-μm aerosol extinction is plotted in Figure 1*a* as a function of altitude between 6 and 25 km. Data has been zonally averaged within the latitude band 20° N to 40° N, the mean tropopause altitude being shown by the black

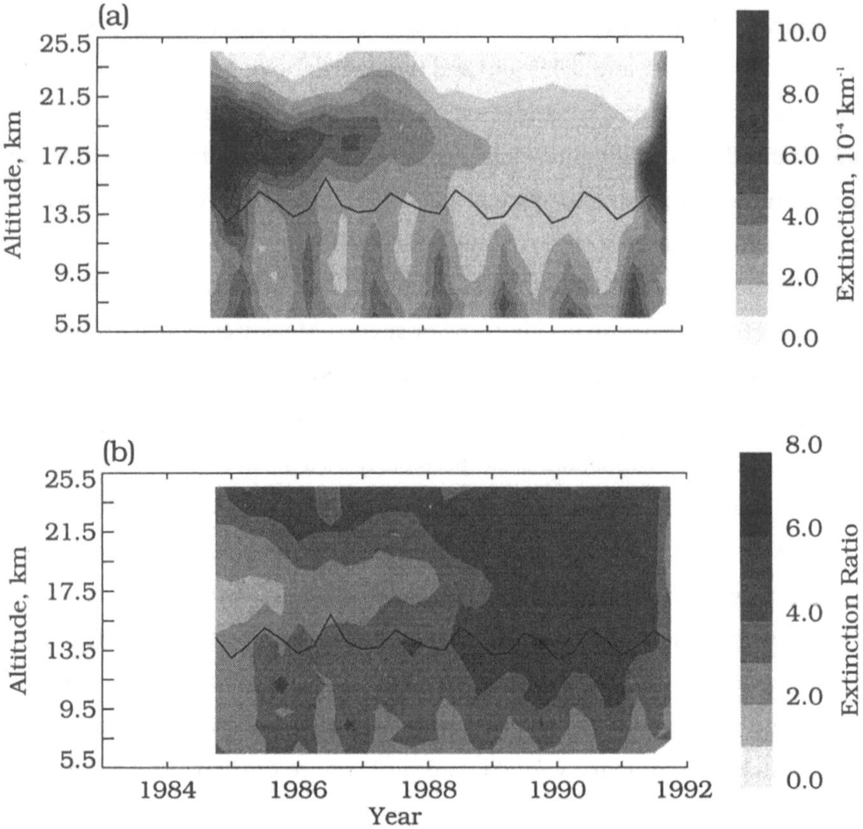

Figure 1. SAGE II zonally averaged data for altitudes from 6.5 km to 24.5 km and the latitude band 20° N to 40° N. The time period covered is October 1984–May 1991, with a resolution of 3 months. The black lines show the location of the tropopause.

(a) The 1.02-µm aerosol extinction. Note the penetration of stratospheric volcanic material into the upper troposphere between 1984 and 1988. Upward extending tongues of extinction, occurring in the spring of each year, are related to the springtime surface aerosol enhancements observed within this latitude band.

(b) The 0.525-µm/1.02-µm aerosol extinction ratio. Note the lower extinction ratios associated with the stratospheric volcanic material and the tropospheric springtime enhancements.

line. Before being plotted, the data has been grouped into 3-month time segments. The time series commences in the fall of 1984 during the decay phase of the aerosol injected into the stratosphere by the eruption of the EL Chichon volcano (McCormick and Veiga, 1992). The final data set shown is for the summer of 1991, just after the eruption of the Pinatubo volcano. A smaller eruption of the Ruiz volcano occurred in November 1985, but the material injected into the stratosphere by this eruption is not readily distinguishable in this figure. Figure 1*a* illustrates the 1.02-μm aerosol extinction, which may be regarded as very approximately proportional to the aerosol mass density. The ratio of the extinction at 0.525 μm to that at 1.02 μm is shown in Figure 1*b*; this ratio is dependent upon the aerosol particle size, larger particles giving rise to smaller ratios and smaller particles to larger ratios. Approximate factors for the conversion of 1.02-μm extinction to mass density and of the 0.525 to 1.02-μm extinction ratio to effective particle radius (Hansen and Hovenier, 1974) are given in Table 1. These have been calculated using a refractive index and mass density corresponding to 75% sulfuric acid (Kent et al., 1995).

The most distinguishing feature of the upper tropospheric aerosol shown in Figure 1*a* is the appearance, in the spring of each year, of tongues of enhanced aerosol extending upwards from the lowest altitude to within a few kilometers of the tropopause. As shown in Figure 1b the extinction ratio decreases at these times, indicating that the particles responsible are larger than those present at other times of year. In the spring of 1985, and to a lesser extent in 1986, there is evidence that aerosol from the large concentration resident in the stratosphere was entering the upper troposphere, where it could be seen coming down to an altitude of about

Table 1. Approximate Values for Conversion from Aerosol Extinction to Mass Density and Effective Particle Radius

(a) Mass density/1-μm Extinction	5 - 15 x 10^{-4} kg.m^{-2}
(b) 0.5-μm Extinction/1.0-μm Extinction	Effective Aerosol Radius
2.0	0.3 μm
3.0	0.2 μm
5.0	0.1 μm

10 km. As shown in Figure 1*b* the size of the stratospheric volcanic aerosol particles from the El Chichon eruption, in 1982, gradually decreases from 1984 to 1988. This reduction is consistent with the expectation that the larger aerosols will settle to the base of the stratosphere where they become lost due to stratospheric-tropospheric exchange processes. Correspondingly, as shown in Figure 1*b*, the size of the volcanic aerosols entering the upper troposphere in 1984–85 was greater than in 1986–88.

Data shown in Figures 2 and 3 are similar to those in Figure 1, except that Figures 2 and 3 each contain six panels, covering the latitude range from 60° S to 60° N. Both of these figures depict a marked latitudinal variation in the behavior of the upper tropospheric aerosol. Referring to the two panels in each figure that depict low latitude data, there is very little penetration of the stratospheric volcanic aerosol into the upper troposphere, even though there are large amounts of aerosol in the stratosphere. Behavior in the latitude band 20° S to 40° S is similar to that already discussed for 20° N to 40° N; an enhancement in aerosol concentration and size is observed in the spring season. This enhancement appears to originate from lower altitudes, the magnitude of the observed changes being less than that of the corresponding effect in the Northern Hemisphere. The panels for the higher latitudes depict more complexity. In the Northern Hemisphere, clearly a seasonal cycle occurs in both the upward and the downward transferred aerosol. This cycle is present in both the concentration and in the aerosol size. The gradual disappearance of volcanic aerosol from the stratosphere and the upper troposphere between 1984 and 1989 is quite clear. In the Southern Hemisphere, the amount of stratospheric aerosol that descends into the upper troposphere is similar to that seen in the Northern Hemisphere. In contrast, the amount of surface-derived material in the upper troposphere in the Southern Hemisphere is much less than in the Northern Hemisphere.

The extinction and extinction ratios shown in Figures 2 and 3 may be converted into approximate mass density and effective radius values, using the conversion factors shown in Table 1. If this conversion is made, the lowest aerosol mass density values in the upper troposphere are approximately 0.1 $\mu gm.m^{-3}$, increasing to

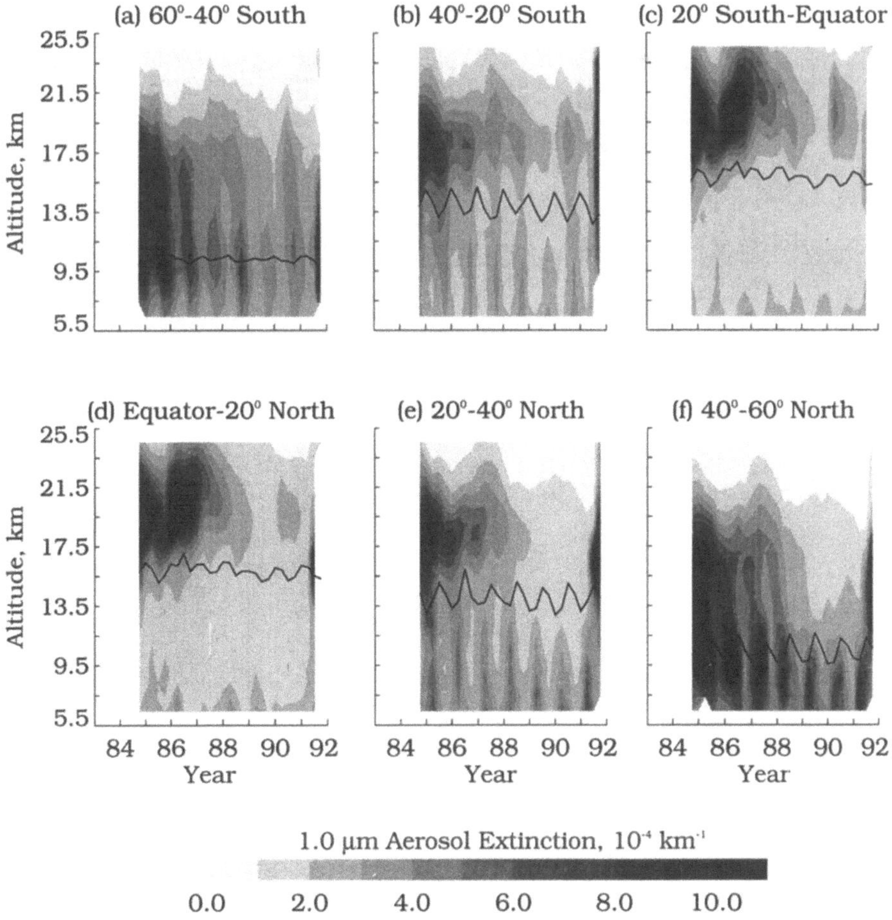

Figure 2. As for Figure 1*a* but data are shown for latitude bands extending from 60° S to 60° N. Each panel shows the zonal mean behavior for a 20° wide latitude band. Note the penetration of stratospheric volcanic material into the upper troposphere at medium and high latitudes but not at low latitudes.

approximately 0.5 μgm.m^{-3} at high latitudes during the volcanically contaminated periods. The observed range for effective radii is from approximately 0.1 to 0.3 μm.

The SAGE II data may also be used to draw maps of the global distribution of aerosol extinction and extinction ratio. An example of such a map is Figure 4, where the global distribution of 1-μm aerosol extinction is plotted for an altitude of 6.5 km.

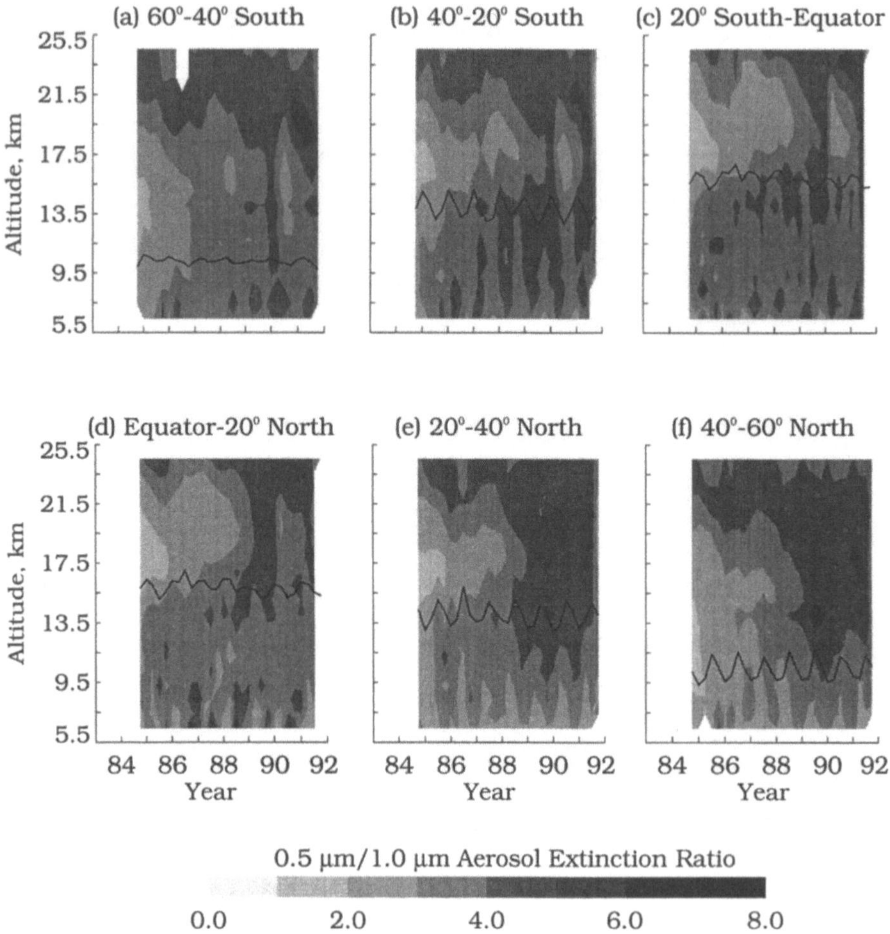

Figure 3. As for Figure 1*b*, but depicting data for latitude bands extending from 60° S to 60° N. In each panel is shown the zonal mean behavior for a 20° wide latitude band.

The map indicates an average for the March-May season only, taken over the years 1985–91. At this altitude, the extinction is dominated by the surface-derived aerosol component, rather than by aerosols entering the troposphere from above. The March–May Northern Hemisphere aerosol enhancement, noted in earlier figures, results in strong asymmetry in the distribution of aerosol extinction between the

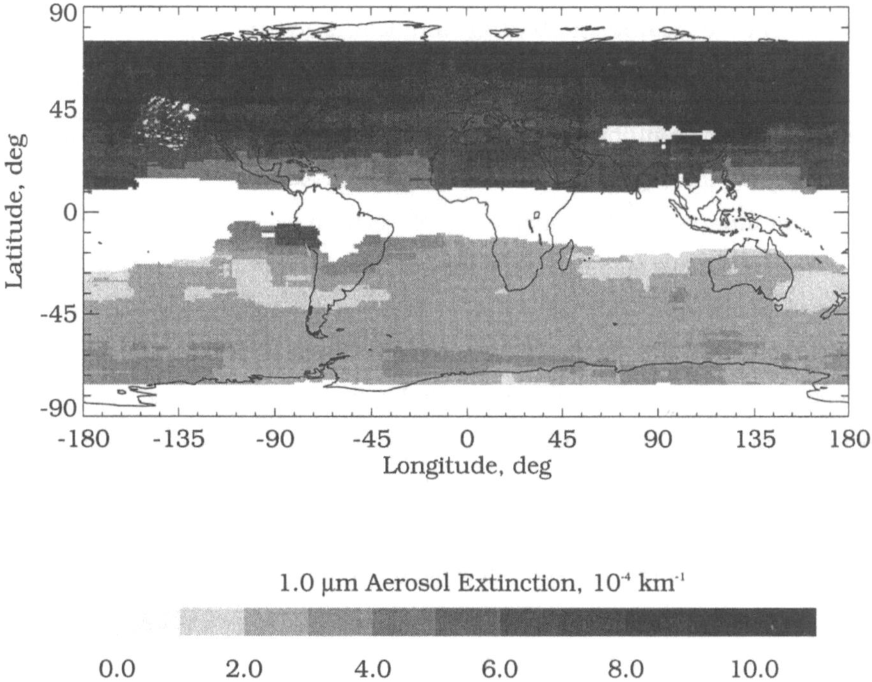

Figure 4. Map depicting the SAGE II 1-μm aerosol extinction for the March–May season, at an altitude of 6.5 km. Data have been averaged over the time period 1985–91. Areas of missing data over the tropics are due to the lack of cloud-free aerosol information for these regions.

Northern and Southern Hemispheres. The figure also shows that even at this altitude, which is the lowest at which two wavelength data are available from SAGE II, the aerosol distribution is essentially zonal in character. Potential aerosol source regions such as the African and Asian deserts and the heavily industrialized area of North America, Europe, and Asia are not outstanding. The absence of data over the tropical regions is attributable to the prevalence of cloud. Even though much of this cloud is thin, SAGE II is unable to obtain a sufficient number of cloud-free data points at an altitude of 6.5 km to complete the map.

Separation of Volcanic and Background Components

In Figures 2 and 3, the difficulty in distinguishing between the volcanic and lower tropospheric contributions to the aerosol in the upper troposphere is apparent, particularly at the higher latitudes. In order to examine the behavior of each component independently, a simple model has been used to separate the effects of each component. This model relies on the fact that the amount of volcanic aerosol in the stratosphere following an eruption is observed to decrease approximately exponentially with time, with a time constant of about 12 months (Yue et al., 1991). The majority of the volcanic effects seen in Figures 2 and 3 are due to the material injected by the eruption of El Chichon, together with smaller contributions from the eruptions of Ruiz and Kelut. The time history of the change in aerosol extinction at 1.02 and 0.525 μm has been examined for each latitude sector and at each 1-km altitude interval from 6 to 25 km. An equation of the form

$$E_t = E_{v0}.exp(t-t_0)/t_d + E_b \qquad (1)$$

where E_t is the aerosol extinction at time t

E_{v0} is the extinction due to the volcanic component at a reference time, t_0

t_d is the decay time constant for the volcanic aerosol

and E_b is a background or baseline extinction level,

has been fitted to each time history curve. The constant t_d has been set to twelve months and E_{v0} and E_b determined for both wavelengths. Care has been taken to avoid time periods in the data where contamination by aerosol from the eruptions of Ruiz and Kelut might affect the fitting process.

The results of separating the volcanic and background aerosol components of the 1.02-μm extinction are shown in Figures 5*a* and 5*b*. In Figure 5*a* is shown the mean zonal distribution of the El Chichon volcanic aerosol for the time period October 1984–August 1985. Similarly, in Figure 5*b* is shown the mean zonal distribution for

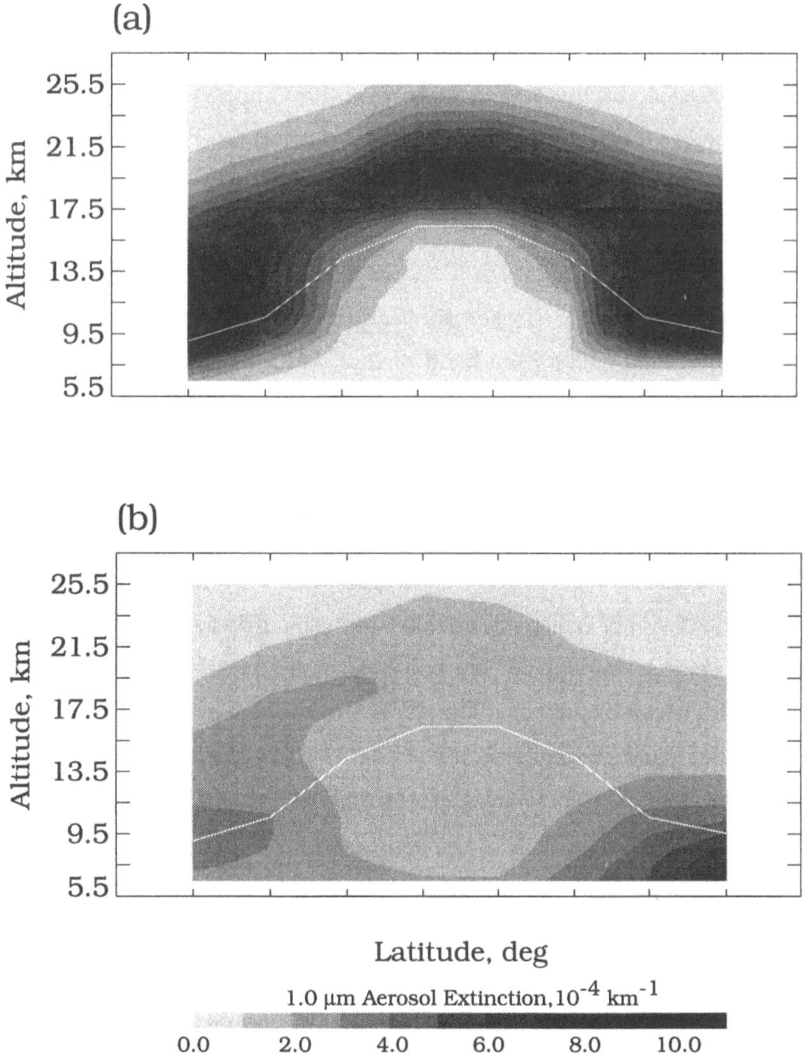

Figure 5. Zonally averaged values of the volcanic and baseline components of the 1.02-μm extinction shown as a function of latitude and altitude. The white lines show the location of the tropopause.

(a) Volcanic component, October 1984–August 1985.

(b) Baseline component.

the background or baseline aerosol. There is assumed to be no long-term variation in the background aerosol, and this plot, which has the seasonal variation removed, does not refer to any particular time period. The baseline aerosol (a term used here in preference to background aerosol, as it is uncertain whether or not the stratospheric aerosol ever reached background levels between 1984 and 1991) has been determined from the lowest mean annual levels observed anytime during the 1984–91 time period. The white lines in Figures 5*a* and 5*b* show the mean tropopause altitudes.

The distributions of the volcanic and baseline aerosols shown in Figures 5*a* and 5*b* are very different. In Figure 5*a* is shown the characteristic banded structure of a well-aged volcanic aerosol. The aerosol banding arises from the combination of two effects (Trepte et al., 1994): the band at low latitudes shows the presence of a reservoir of volcanic material over the tropics, while those at high latitudes are caused by the departure of the tropopause from isentropic surfaces, which permits the accumulation of the aerosol following transport from low latitudes. The most interesting feature of the distribution of volcanic aerosol is the relationship between the contours of aerosol extinction and the tropopause. Between latitudes of 30° S and 30° N the contours are effectively parallel and very little aerosol is observed below the tropopause. Polewards of 30° the extinction contours descend several kilometers relative to the mean tropopause. The upper troposphere close to the tropopause at these latitudes yield extinction levels similar to those in the stratosphere, and the tropopause, at least in its annual mean form, does not resemble a material boundary. The distribution shown in Figure 5*a* has been used to calculate the approximate mass of volcanic aerosol lying below the tropopause as a function of latitude. The results of this calculation are shown in Table 2, where the amount within each latitude band has been expressed as a percentage of the global total (stratospheric plus tropospheric). The tropospheric latitudinal distribution peaks within the 40° to 60° latitude band in each hemisphere. The total aerosol mass integrated over all latitudes for the upper troposphere is about 14% of the global total.

As shown in Figure 5*b* the distribution of the 1.02-μm extinction for the baseline aerosol is quite different from that for the volcanic aerosol. In the Northern Hemisphere, a maximum occurs with the extinction increasing toward higher latitudes. The equatorial region is relatively clean, although the extinction is not as low as that of the volcanic component. At high southern latitudes a curious

Table 2. Estimated Percentage of Global Volcanic Aerosol Mass Lying Within the Upper Troposphere—Latitudinal Variation

Latitude Band	Upper Troposphere Mass Percentage
90° S–60° S	1.7
60° S–40° S	2.4
40° S–20° S	1.8
20° S–Eq	1.0
Eq–20° N	1.0
20° N–40° N	1.9
40° N–60° N	3.0
60° N–90°N	1.3
----------------------	------
Global total	14.0

maximum occurs near the tropopause. The cause of this is not clear, although examination of data obtained over several years shows that it is related to the wintertime processing of aerosol within the South polar vortex (Thomason and Poole, 1993). The enhanced extinction observed in the stratosphere, poleward of 30° S, is believed not to be a true feature of the background aerosol but to be an artifact caused by a residual quantity of stratospheric aerosol.

Wavelength extinction ratios corresponding to the distributions in Figures 5*a* and 5*b* have been calculated. The principal feature of those for the volcanic component is a strong gradient with altitude in the stratosphere in which high ratios (smaller particle radii) occur at the upper levels and lower ratios (larger particle radii) occur close to the tropopause. This behavior is as previously documented for the stratosphere, where particle size decreases with altitude (e.g., Brogniez and Lenoble, 1988). Values for extinction ratio of the volcanic component within the troposphere are subject to greater error than those in the stratosphere due to the small volcanic signature relative to the baseline component. In general, the sizes of the volcanic aerosol particles are large (effective radii 0.25 to 0.5 μm), similar to those observed in the lower stratosphere. Wavelength extinction ratios for the baseline aerosol component are larger than those for the volcanic component, except at the higher stratospheric altitudes, indicating smaller particle sizes (effective radii 0.1 to 0.3 μm) in the upper troposphere.

Seasonal variations are observed in all the extinction values described. These can provide useful insights into the sources and mechanisms controlling the aerosol in the upper troposphere. Figures *6a* and *6b* are examples of seasonal differences that depict, respectively, the wavelength extinction ratio for the baseline aerosol during two seasons, March–May and September–November. Both distributions are different from that shown in Figure 5*b*. In particular, no indication is given in eitherFigure *6a* or Figure *6b* of the Northern Hemisphere maximum shown in Figure 5*b*. A feature apparent in both Figures *6a* and *6b* is the appearance of a minimum in extinction ratio at an altitude of 6 km and latitude of 30° N (Figures *6a* and *6b*) and 30° S (Figure *6b*). Both minima in the extinction ratio correspond to a maximum in particle size and indicate the appearance, on a seasonal basis at these latitudes and altitude, of surface derived large particles. In both hemispheres the peak occurrence is in local spring and that in the Northern Hemisphere is of greater magnitude than its equivalent in the Southern Hemisphere. The difference in latitudinal behavior between these distributions and that in Figure 5*b*, which also shows a seasonal cycle with a springtime maximum in the Northern Hemisphere, indicates that most likely different aerosol phenomena are being observed. The phenomena at 30° latitude in either hemisphere have the characteristics of surface-derived dust in terms of both latitude of origin and particle size. The maximum at high northern latitudes corresponds more closely to an aerosol from an anthropogenic source in terms of its latitude, smaller particle size, and deeper atmospheric distribution.

Seasonal variations also take place in the amount of penetration of stratospheric volcanic aerosol into the upper troposphere at middle and high latitudes. The relative masses of volcanic aerosol observed within latitude bands of 60° S to 40° S and 40° N to 60° N are shown as a function of season in Table 3. These latitude bandshave been chosen to illustrate the seasonal changes as those in which the greatest stratospheric-tropospheric aerosol mass transfer occurred. Both hemispheres show a similar, although not identical, seasonal dependence. Minimum downward transfer is observed in fall and winter and maximum in spring and summer. The peak-to-peak variation is somewhat less than a factor of 2.

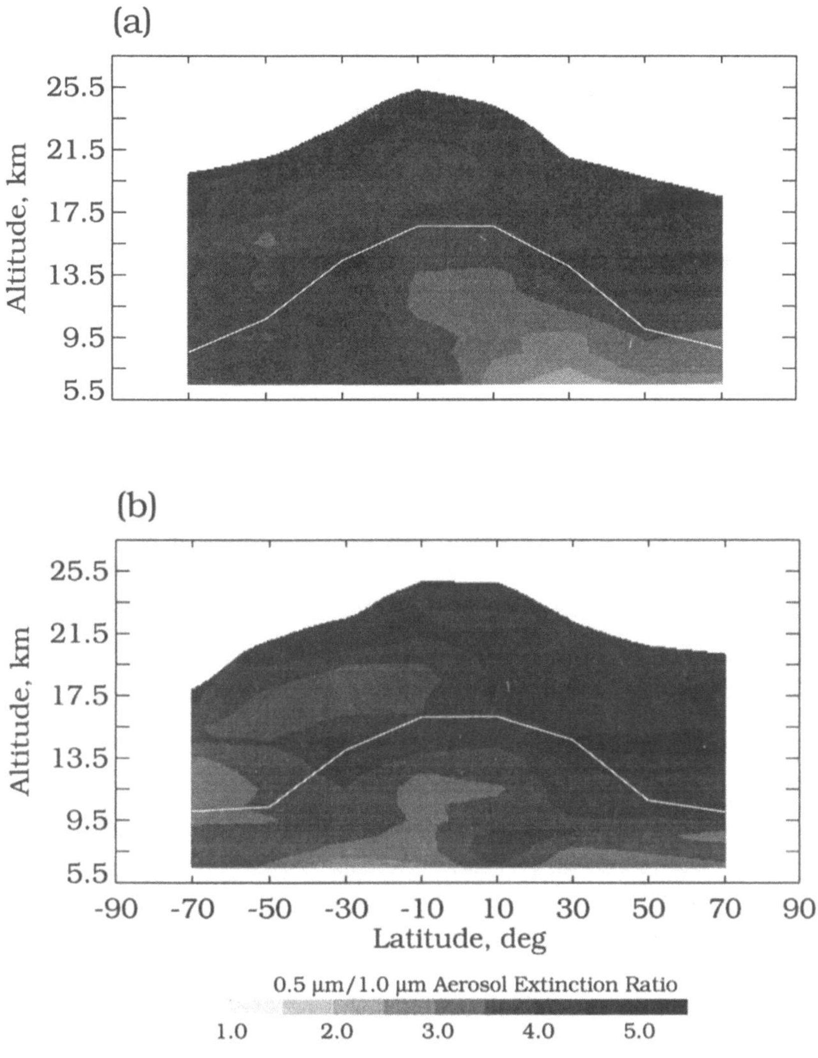

Figure 6. Seasonal characteristics of the zonally averaged 0.525- and 1.02-μm aerosol extinction ratios for the baseline aerosol component. The white lines show the location of the tropopause.

(a) March–May.

(b) September–November.

Table 3. Estimated Percentage of Global Volcanic Aerosol Mass Lying Within the Upper Troposphere—Seasonal Variation

Season	Upper Troposphere Mass Percentage	
	60° S–40° S	40° N–60° N
March–May	2.2	3.0
June–August	1.6	3.9
September–November	2.9	2.3
December–February	2.6	2.4

Summary and Discussion

The most significant feature of the baseline aerosol in the upper troposphere is the strong north-south asymmetry. This was previously noted in the SAGE II data set by Kent et al. (1991) and is clearly shown in Figure 4. It is similar to that seen by Durkee et al. (1991) in data obtained by the NOAA-7 Advanced Very High Resolution Radiometer (AVHRR). The AVHRR data are presented as maps of aerosol optical depth and, as an integrated product, are strongly weighted towards the lowest region of the atmosphere. These data are also only available for oceanic regions. Nevertheless, they show considerable similarity to the SAGE II global aerosol distribution, the optical depth in the Northern Hemisphere being greater than that in the Southern Hemisphere by a factor of the order of 1.5. This variation is attributed to the hemispheric difference in continental aerosol sources both natural and anthropogenic. Plumes of particles are observed originating from continental industrial centers and from arid desert regions. The low latitude aerosol extinction minimum seen in the SAGE II data set is qualitatively similar to that observed by Clarke (1993). On a horizontal transit flight from Darwin to Tokyo, at an altitude of 10 km, very low mass concentrations of optically active nucleii were observed when close to the Equator.

The subtropical maxima in the baseline tropospheric aerosol wavelength extinction ratio, and particle size, are consistent with the theory that these aerosols are derived from the earth's desert regions. It has been known for some time that long-range transport of dust occurs from both the Saharan and Asian deserts (Schütz, 1980; Uematsu et al., 1983; Prospero and Nees, 1977). This dust can be lofted to altitudes of 6 to 10 km (Prospero and Carlson, 1972; Bryson and Barreis, 1967), and transported over several thousand kilometers (Shaw, 1980). The dust has a springtime maximum, and particle sizes are observed to be larger than those of the background aerosol (Duce et al., 1980; Shaw, 1980). Deserts in the Southern Hemisphere are less extensive than those in the Northern Hemisphere but long-range transport of dust has been observed from both the southern African and Australian deserts (Gillette et al., 1993). The lesser aerosol particle size maximum for southern subtropical latitudes, as opposed to the northern, which is observed in the SAGE II data, is consistent with the global distribution of deserts.

The maximum in 1.02-μm aerosol extinction at high northern latitudes, as shown in Figure 5*b*, is most obviously interpreted in terms of the industrial hazes that are seen in these regions. These have been studied extensively in the form of arctic haze (Barrie, 1986) and are generally viewed as a wintertime phenomenon. This seasonal variation is in distinction to that observed in the SAGE II data, where a maximum is observed in spring, and aerosol extinction values in the other seasons are found to be fairly constant. The seasonal variation reported for arctic haze is typical for altitudes in, or close to, the boundary layer where a minimum is observed in summer, attributed to scavenging by stratus clouds (Radke et al., 1989). The seasonal behavior of the aerosol at the higher altitudes may be considerably different. The haze is extensively layered up to several kilometers. Radke et al. (1984) and Heintzenberg et al. (1991) have shown that these layers exist (up to at least 7-km altitude) throughout the summer season. The optical properties of these layers were found to be relatively homogeneous on a ferry flight from Munich to Spitsbergen, indicating long range summertime transport of aerosol from midlatitude industrial regions to the pole.

The distribution shown in Figure 5*a* for the volcanic component of the aerosol is consistent with current models for transport in the upper troposphere and lower stratosphere. In these models upward motion takes place primarily in the tropical regions and is associated with deep convective activity (World Meteorological

Organization, 1985; Andrews et al., 1987). Return flow is primarily a midlatitude phenomenon and actual exchange takes place by means of tropopause folding, the latter occurring in conjunction with frontal development and jet streaks (Carlson, 1991). This picture is in agreement with the distribution of volcanic aerosol shown in Figure 5*a*. In this figure, which indicates the mean annual behavior of the volcanic aerosol component, relatively little aerosol is depicted below the tropopause between latitudes 30° S and 30° N, a region of upward motion. Polewards of these latitudes the uppermost layers of the troposphere have aerosol concentrations similar to those in the stratosphere, indicative of strong downward movement of stratospheric material. As shown in Table 2, the total mass of volcanic aerosol in the upper troposphere reaches a maximum in the 40° to 60° latitude bands, although the mass density shows little change between middle and high latitudes. As shown in Table 3 the amount of volcanic aerosol in the upper troposphere peaks in local spring and summer; this phenomenon is somewhat at variance with the seasonal variation of tropopause folding, which reaches its peak in winter and spring. Observations of stratospheric air entering the upper troposphere in middle and high northern latitudes have been made using airborne and ground-based lidar. The CO_2 lidar measurements made at Boulder, Colorado (Post, 1986; Post et al., 1993; M.J. Post, personal communication, 1994) have shown clearly volcanic aerosols entering the upper troposphere following the eruptions of the El Chichon and Pinatubo volcanoes. The most recent measurements show the stratospheric aerosol from the eruption of Pinatubo entering the upper troposphere preferentially in spring and summer, visible as an enhancement above background down to an altitude of 7 to 8 km. Browell et al. (1994) have used airborne lidar to study intrusions of stratospheric ozone and aerosol into the upper troposphere at high latitudes. These were found on more than 40% of flights made in the summer of 1990, the intrusions being visible down to low altitudes.

The SAGE II data may be used to estimate a residence time for aerosol entering the upper troposphere from above. The exponential decay time for the mass of volcanic aerosol in the stratosphere is approximately 12 months. An amount equal to the average upper tropospheric loading would enter the troposphere in about

2 months, yielding an approximate figure for the mean upper tropospheric residence time. This simple calculation takes no account of seasonal variations, but a residence time of 2 months is in reasonable agreement with that previously estimated from CO_2 lidar measurements. The time period is much longer than that associated with aerosols in the lower troposphere but agrees with other estimates for the lifetime of aerosols in the upper troposphere (Jaenicke, 1988).

In the above discussion, the relative freedom of the low latitude upper troposphere from volcanic aerosol has been attributed solely to dynamic forces; the true situation is not that simple. Aerosols will descend to lower altitudes owing to gravitational sedimentation, and this sedimentation is most likely responsible for the small amounts of volcanic aerosol that are seen in the upper tropical troposphere. Other aerosol may be brought down to lower altitudes in the descending regions of the Walker circulation. Neither mechanism can account for the steep gradient in aerosol extinction that occurs between an altitude just above the mean tropical tropopause and about 14 km. The tropics are a region of deep convective activity, tropical convective complexes being observed to reach to altitudes of 17 to 18 km (Ackerman et al., 1988). Scavenging of aerosols due to entrainment in high cloud will take place, producing a relatively clean upper troposphere at low latitudes (Clarke, 1993). It is thus possible that a somewhat greater amount of volcanic aerosol is entering the tropical upper troposphere than is implied by the distribution shown in Figure 5*a*. The true aerosol balance is determined by a combination of dynamic and gravitational forces as well by a consideration of the different tropospheric removal processes.

Acknowledgments

The author wishes to thank the many individuals at NASA Langley, Science and Technology Corporation, and Science Applications International Corporation, who have been involved in the processing of the data used. He is grateful to J. Theon, R. Kakar, and J. Kaye of NASA Headquarters, Washington, D.C., for their interest in and support of this work. The work has been supported under NASA contracts NAS1-19603 and NAS1-19976.

References

Ackerman TP, Liou K-N, Valero FJP, Pfister L (1988) Heating rates in tropical anvils. J Atmos Sci 45:1606-1623

Andrews DG, Holton JR, Leovy CB (1987) Middle atmosphere dynamics. Academic Press, New York

Barrie LA, (1986) Arctic air pollution: An overview of current knowledge. Atmos Environ 20:643-663

Brogniez C, Lenoble J (1988) Size distribution of stratospheric aerosols from SAGE II multiwavelength extinctions. In Aerosols and Climate, eds. PV Hobbs and MP McCormick, A Deepak Publishing, Hampton, Virginia

Browell EV, Fenn MA, Butler CF, Grant WB, Harris RC, Shipham MC (1994) Ozone and aerosol distributions in the summertime troposphere over Canada. J Geophys Res 99:1739-1755

Bryson RA, Barreis DA (1967) Possibilities of major climatic modification and their implication: Northewest India, a case study. Bull Am Meteorol Soc, 48:136

Carlson TN, (1991) Mid-latitude weather systems, Harper Collins, London

Chu WP, McCormick MP, Lenoble J, Brogniez C, Pruvost P (1989) SAGE II inversion algorithm. J Geophys Res 94:8339-8351

Clarke AD (1993) Atmospheric nuclei in the Pacific midtroposphere: Their nature, concentration and evolution. J Geophys Res 98:20633-20647

Duce RA, Unni CK, Ray BJ, Prospero JM, Merrill JT (1980) Long-range atmospheric transport of soil dust from Asia to the tropical North Pacific: Temporal variability. Science 209:1522-1524

Durkee PA, Pfeil F, Frost E, Shema R (1991) Global analysis of aerosol particle characteristics. Atmos Environ 25A: 2457-2471

Gillette DA, Patterson EM Jr, Prospero JM, Jackson ML (1993) Soil aerosols. In Aerosol Effects on Climate, ed. SG Jennings, University of Arizona Press, Tucson, 73-109

Hansen JE, Hovenier JW (1974) Interpretation of the polarisation of Venus. J Atmos Sci 31:1137-1160

Heintzenberg J, Ström J, Ogren JA (1991) Vertical profiles of aerosol properties in the summer troposphere of central Europe, Scandinavia and the Svalbard region. Atmos Environ 25A:621-627

Hofmann DJ (1993) 20 years of balloonborne tropospheric aerosol measurements at Laramie, Wyoming. J Geophys Res 98:12753

Jaenicke R (1988) Aerosol physics and chemistry. In Landolt-Boerstein, Numerical Data and Functional Relationships in Science and Technology, V, 4b (Meteorology) Chapter 9, 301-457

Kent GS, McCormick MP (1991) Separation of cloud and aerosol in two-wavelength satellite occulatation data. Geophys Res Lett 18:428-431

Kent GS, Farrukh UO, Wang PH, Deepak A (1988) SAGE I and SAM II measurements of 1.0 μm aerosol extinction in the free troposphere. J Appl Meteorol 27:269-279

Kent GS, Schaffner SK (1989) Comparison of 1 μm satellite aerosol extinction with CO_2 lidar backscatter. SPIE Proceedings 1181:252-259

Kent GS, McCormick MP, Schaffner SK (1991) Global optical climatology of the free tropospheric aerosol from 1.0 μm satellite occultation measurements. J Geophys Res 96:5249-5267.

Kent GS, Winker DM, Osborn MT, McCormick MP, Skeens KM (1992) A model for the separation of cloud and aerosol in SAGE II occultation data. J Geophys Res 98:20725-20735

Kent GS, Wang P-H, McCormick MP, Skeens KM (1995) Multiyear SAGE II measurements of upper tropsopheric aerosol characteristics. J Geophys Res (in press)

McCormick MP (1992) SAGE II: An overview. Adv Space Res 7:(3)219-(3)226.

McCormick MP, Veiga RE (1992) SAGE II measurements of early Pinatubo aerosols. Geophys Res Lett 19:155-158

Post MJ (1986) Atmospheric purging of El Chichon debris. J Geophys Res 91:5222-5228

Post MJ, Grund CJ, Langford AO, Proffitt MH (1992) Observations of Pinatubo ejecta over Boulder, Colorado, by lidars of three different wavelengths. Geophys Res Lett 19:195-198

Prospero JM, Carlson TM, Vertical and areal distribution of Saharan dust over the western North Atlantic Ocean. J Geophys Res 77:5255-5265

Prospero JM, Nees RT (1977) Dust concentration in the atmosphere of the equatorial North Atlantic: Possible relationship to the Sahelian drought. Science 196:1196-1198.

Pueschel RF, Livingston JM, Ferry GV, DeFelice TE (1994) Aerosol abundances and optical characterisitics in the Pacific basin free troposphere. Atmos Environ 28:991-960

Radke LF, Lyons JH, Hegg DA, Hobbs PV, Bailey IH (1984) Airborne observations of Arctic aerosols. I: Characteristics of Arctic haze. Geophys Res Lett 11:393-396

Radke LF, Brock CA, Lyons JH, Hobbs PV, Schnell RC (1989) Aerosol and lidar measurements of hazes in mid-latitude and polar airmasses. Atmos Environ 23:2417-2430

Schütz L (1980) Long range transport of desert dust with special emphasis on the Sahara. Ann NY Acad Sci 77:515-531

Shaw GE (1980) Transport of Asian desert dust to the Hawaiian Islands. J Appl Meteorol 19:1254-1259

Thomason LW, Poole LR (1993) On the use of Antarctic stratospheric aerosol properties as diagnostics of vortex processes. J Geophys Res 98:23003-23013

Trepte CR, Thomason LW, Kent GS (1994) Banded structures in stratospheric aerosol distributions. Geophys Res Lett 21:2397-2400

Uematsu M, Duce RA, Prospero JM, Chen L, Merrill JT, McDonald RL (1983) Transport of mineral aerosol from Asia over the North Pacific Ocean. J Geophys Res 88:5343-5352

Yue GK, McCormick MP, Chiou EW (1991) Stratospheric aerosol optical depth observed by the Stratospheric Aerosol and Gas Experiment II: Decay of the El Chichon and Ruiz volcanic perturbations. J Geophys Res 96:5209-5219

World Meteorological Organization (1985) Atmospheric Ozone 1985. Report No. 16, Geneva, Switzerland 1:151-240

List of Participants

Dr. Marcia Baker
Geophysics Program, AK-50
University of Washington
Seattle, WA 98195
USA

Dr. Lennart Bengtsson
Max-Planck Institute for Meteorology
D-20146 Hamburg 13
GERMANY

Dr. Andre Berger
Institut d'Astronomie et de Geophysique G. Lemaitre
Universite Catholique de Louvain
B-1348 Louvain-la-Nueve
BELGIUM

Dr. Stephen Borrmann
Institute for Atmospheric Physics
University of Mainz
D-55099 Mainz
GERMANY

Dr. Robert Charlson
Department of Atmospheric Sciences
University of Washington
Seattle, WA 98195
USA

Dr. William Collins
Center for Clouds, Chemistry and Climate
Scripps Institution of Oceanography
University of California, San Diego
La Jolla, CA 92093-0239
USA

Dr. Paul J. Crutzen
Max-Planck Institute for Chemistry
D-55020 Mainz
GERMANY

Dr. Kerry Emanuel
Earth, Atmospheric, & Planetary Sciences
Massachusetts Institute of Technology
Cambridge, MA 02139
USA

Dr. Yves Fouquart
Laboratoire d'Optique Atmospherique
Unviersite des Sciences et Techniques de Lille
F-59655 Villeneuve d'Ascq
FRANCE

Dr. Hartmut Grassl
Max-Planck Institute for Meteorology
20146 Hamburg
GERMANY

Dr. Robert Grossman
APAS Department, #391
University of Colorado
Boulder, CO 80309
USA

Dr. Claus Hammer
Geophysical Department
University of Copenhagen
DK-2200 Copenhagen N.
DENMARK

Dr. Jost Heintzenberg
Institute for Tropospheric Research
D-04303 Leipzig
GERMANY

Dr. Andrew Heymsfield
National Center for Atmospheric Research
Boulder, CO 80307
USA

Dr. Jim Holton
Atmospheric Science AK-40
University of Washington
Seattle, WA 98195
USA

Dr. Jean Jouzel
Laboratoire de Modélisation du Climat et de l'Environnement, CEA/DSM
CE Sarclay 91191 Gif/Yvette Cedex
FRANCE

Dr. Geoff Kent
Science and Technology Corporation
Hampton, VA 23669
USA

Dr. Dieter Kley
Institute for Chemistry and Dynamics of the Geosphere
Forschungszentrum Juelich
GERMANY

Dr. Joach Kuettner
National Center for Atmospheric Research
Boulder, CO 80307
USA

Dr. Joseph Lelieveld
Air Quality Department
Wageningen Agricultural University
6700 EV Wageningen
THE NETHERLANDS

Dr. Herve Le Treut
Laboratoire de Meteorologie Dynamique
Ecole Normale Superieure
75231 Paris Cedex 05
FRANCE

H. V. Nguyen
Center for Clouds, Chemistry and Climate
Scripps Institution of Oceanography
University of California, San Diego
La Jolla, CA 92093-0239
USA

Dr. Thor Erik Nordeng
European Centre for Medium Range Weather Forecasts
Shinfield Park, Reading
RG2 9AX
ENGLAND

Dr. Thomas Peter
Max-Planck Institute for Chemistry
D-55020 Mainz
GERMANY

Dr. Raymond Pierrehumbert
Department of Geophysical Sciences
University of Chicago
Chicago, IL 60637
USA

Dr. Hans Pruppacher
Institute for Atmospheric Physics
University of Mainz
D-55099 Mainz
GERMANY

Dr. V. Ramanathan
Center for Clouds, Chemistry and Climate
Scripps Institution of Oceanography
University of California, San Diego
La Jolla, CA 92093-0239
USA

Dr. Phil Rasch
National Center for Atmospheric Research
Boulder, CO 80307
USA

Dr. Erich Roeckner
Max-Planck Institute for Meteorology
20146 Hamburg 13
GERMANY

Dr. Robert Sadourny
Laboratoire de Meteorologie Dynamique
Ecole Normale Superieure
74231 Paris Cedex 05
FRANCE

Dr. Stephen Schwartz
Environmental Chemistry Division
Brookhaven National Laboratory
Upton, NY11973
USA

Dr. Anthony Slingo
Hadley Centre for Climate Prediction and Research
Berkshire RG12 2SY
UNITED KINGDOM

The ASI Series Books Published as a Result of Activities of the Special Programme on Global Environmental Change

This book contains the proceedings of a NATO Advanced Research Workshop held within the activities of the NATO Special Programme on Global Environmental Change, which started in 1991 under the auspices of the NATO Science Committee.

The volumes published as a result of the activities of the Special Programme are:

Vol. 1: **Global Environmental Change.**
Edited by R. W. Corell and P. A. Anderson. 1991.

Vol. 2: **The Last Deglaciation: Absolute and Radiocarbon Chronologies.**
Edited by E. Bard and W. S. Broecker. 1992.

Vol. 3: **Start of a Glacial.** Edited by G. J. Kukla and E. Went. 1992.

Vol. 4: **Interactions of C, N, P and S Biogeochemical Cycles and Global Change.**
Edited by R. Wollast, F. T. Mackenzie and L. Chou. 1993.

Vol. 5: **Energy and Water Cycles in the Climate System.**
Edited by E. Raschke and D. Jacob. 1993.

Vol. 6: **Prediction of Interannual Climate Variations.**
Edited by J. Shukla. 1993.

Vol. 7: **The Tropospheric Chemistry of Ozone in the Polar Regions.**
Edited by H. Niki and K. H. Becker. 1993.

Vol. 8: **The Role of the Stratosphere in Global Change.**
Edited by M.-L. Chanin. 1993.

Vol. 9: **High Spectral Resolution Infrared Remote Sensing for Earth's Weather and Climate Studies.**
Edited by A. Chedin, M.T. Chahine and N.A. Scott. 1993.

Vol. 10: **Towards a Model of Ocean Biogeochemical Processes.**
Edited by G. T. Evans and M.J.R. Fasham. 1993.

Vol. 11: **Modelling Oceanic Climate Interactions.**
Edited by J. Willebrand and D.L.T. Anderson. 1993.

Vol. 12: **Ice in the Climate System.** Edited by W. Richard Peltier. 1993.

Vol. 13: **Atmospheric Methane: Sources, Sinks, and Role in Global Change.**
Edited by M. A. K. Khalil. 1993.

Vol. 14: **The Role of Regional Organizations in the Context of Climate Change.**
Edited by M. H. Glantz. 1993.

Vol. 15: **The Global Carbon Cycle.**
Edited by M. Heimann. 1993.

Vol. 16: **Interacting Stresses on Plants in a Changing Climate.**
Edited by M. B. Jackson and C. R. Black. 1993.

Vol. 17: **Carbon Cycling in the Glacial Ocean: Constraints on the Ocean's Role in Global Change.**
Edited by R. Zahn, T. F. Pedersen, M. A. Kaminski and L. Labeyrie. 1994.

Vol. 18: **Stratospheric Ozone Depletion/UV-B Radiation in the Biosphere.**
Edited by R. H. Biggs and M. E. B. Joyner. 1994.

Vol. 19: **Data Assimilation: Tools for Modelling the Ocean in a Global Change Perspective.**
Edited by P. O. Brasseur and J. Nihoul. 1994.

Vol. 20: **Biodiversity, Temperate Ecosystems, and Global Change.**
Edited by T. J. B. Boyle and C. E. B. Boyle. 1994.

Vol. 21: **Low-Temperature Chemistry of the Atmosphere.**
Edited by G. K. Moortgat, A. J. Barnes, G. Le Bras and J. R. Sodeau. 1994.

Vol. 22: **Long-Term Climatic Variations – Data and Modelling.**
Edited by J.-C. Duplessy and M.-T. Spyridakis. 1994.

Vol. 23: **Soil Responses to Climate Change.**
Edited by M. D. A. Rounsevell and P. J. Loveland. 1994.

Vol. 24: **Remote Sensing and Global Climate Change.**
Edited by R. A. Vaughan and A. P. Cracknell. 1994.

Vol. 25: **The Solar Engine and Its Influence on Terrestrial Atmosphere and Climate.**
Edited by E. Nesme-Ribes. 1994.

Vol. 26: **Global Precipitations and Climate Change.**
Edited by M. Desbois and F. Désalmand. 1994.

Vol. 27: **Cenozoic Plants and Climates of the Arctic.**
Edited by M. C. Boulter and H. C. Fisher. 1994.

Vol. 28: **Evaluating and Monitoring the Health of Large-Scale Ecosystems.**
Edited by D. J. Rapport, C. L. Gaudet and P. Calow. 1995.

Vol. 29: **Global Environmental Change Science: Education and Training.**
Edited by D. J. Waddington. 1995.

Vol. 30: **Ice Core Studies of Global Biogeochemical Cycles.**
Edited by R. J. Delmas. 1995.

Vol. 31: **The Role of Water and the Hydrological Cycle in Global Change.**
Edited by H. R. Oliver and S. A. Oliver. 1995.

Vol. 32: **Atmospheric Ozone as a Climate Gas.**
Edited by W.-C. Wang and I. S. A. Isaksen. 1995.

Vol. 33: **Carbon Sequestration in the Biosphere.**
Edited by M. A. Beran. 1995.

Vol. 34: **Climate Sensitivity to Radiative Perturbations: Physical Mechanisms and Their Validation.**
Edited by H. Le Treut. 1996

Vol. 35: **Clouds, Chemistry and Climate.**
Edited by P. J. Crutzen and V. Ramanathan. 1996.

Vol. 36: **Diachronic Climatic Impacts on Water Resources.**
Edited by Andreas N. Angelakis and Arie S. Issar. 1996.

Vol. 37: **Climate Change and World Food Security.**
Edited by Thomas E. Downing. 1996.

MIX
Papier aus verantwortungsvollen Quellen
Paper from responsible sources
FSC® C105338

If you have any concerns about our products,
you can contact us on
ProductSafety@springernature.com

In case Publisher is established outside the EU,
the EU authorized representative is:
Springer Nature Customer Service Center GmbH
Europaplatz 3, 69115 Heidelberg, Germany

Printed by Libri Plureos GmbH
in Hamburg, Germany